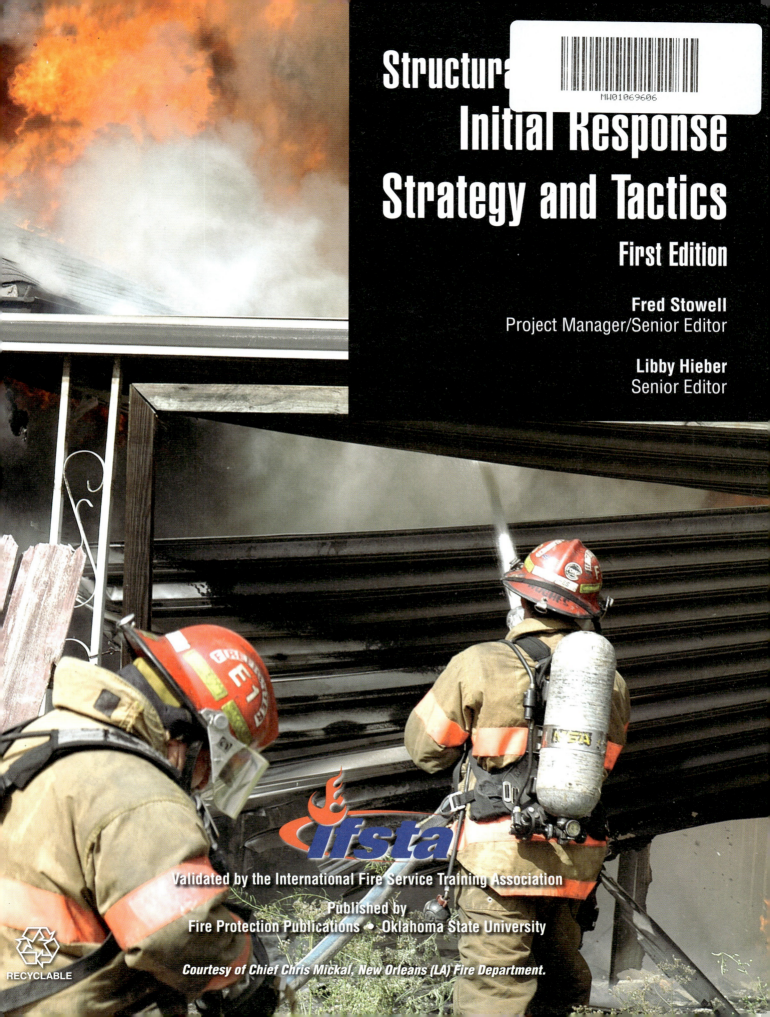

Structural Fire Fighting: Initial Response Strategy and Tactics

First Edition

Fred Stowell
Project Manager/Senior Editor

Libby Hieber
Senior Editor

Validated by the International Fire Service Training Association

Published by
Fire Protection Publications • Oklahoma State University

Courtesy of Chief Chris Mickal, New Orleans (LA) Fire Department.

The International Fire Service Training Association

The International Fire Service Training Association (IFSTA) was established in 1934 as a *nonprofit educational association of fire fighting personnel who are dedicated to upgrading fire fighting techniques and safety through training*. To carry out the mission of IFSTA, Fire Protection Publications was established as an entity of Oklahoma State University. Fire Protection Publications' primary function is to publish and disseminate training texts as proposed and validated by IFSTA. As a secondary function, Fire Protection Publications researches, acquires, produces, and markets high-quality learning and teaching aids as consistent with IFSTA's mission.

The IFSTA Validation Conference is held the second full week in July. Committees of technical experts meet and work at the conference addressing the current standards of the National Fire Protection Association® and other standard-making groups as applicable. The Validation Conference brings together individuals from several related and allied fields, such as:

- Key fire department executives and training officers
- Educators from colleges and universities
- Representatives from governmental agencies
- Delegates of firefighter associations and industrial organizations

Committee members are not paid nor are they reimbursed for their expenses by IFSTA or Fire Protection Publications. They participate because of commitment to the fire service and its future through training. Being on a committee is prestigious in the fire service community, and committee members are acknowledged leaders in their fields. This unique feature provides a close relationship between the International Fire Service Training Association and fire protection agencies, which helps to correlate the efforts of all concerned.

IFSTA manuals are now the official teaching texts of most of the states and provinces of North America. Additionally, numerous U.S. and Canadian government agencies as well as other English-speaking countries have officially accepted the IFSTA manuals.

Copyright © 2010 by the Board of Regents, Oklahoma State University

All rights reserved. No part of this publication may be reproduced in any form without prior written permission from the publisher.

ISBN 978-0-87939-395-3   Library of Congress Control Number: 2010935224

First Edition, First Printing, October 2010   *Printed in the United States of America*

10   9   8   7   6   5   4   3   2   1

If you need additional information concerning the International Fire Service Training Association (IFSTA) or Fire Protection Publications, contact:

Customer Service, Fire Protection Publications, Oklahoma State University
930 North Willis, Stillwater, OK 74078-8045
800-654-4055   Fax: 405-744-8204

For assistance with training materials, to recommend material for inclusion in an IFSTA manual, or to ask questions or comment on manual content, contact:

Editorial Department, Fire Protection Publications, Oklahoma State University
930 North Willis, Stillwater, OK 74078-8045
405-744-4111   Fax: 405-744-4112   E-mail: editors@osufpp.org

---

Oklahoma State University in compliance with Title VI of the Civil Rights Act of 1964 and Title IX of the Educational Amendments of 1972 (Higher Education Act) does not discriminate on the basis of race, color, national origin or sex in any of its policies, practices or procedures. This provision includes but is not limited to admissions, employment, financial aid and educational services.

Cover photo courtesy of Tom Aurnhammer.

Chapter Summary

Chapters
1. Managing the Incident 9
2. Fire Behavior, Building Construction, and Building Occupancy 43
3. Size-Up: Evaluation and Assessment 93
4. Strategy 141
5. Tactics 161
6. Residential Scenarios 189
7. Commercial Scenarios 259
8. Special Hazards Scenarios 311

Appendices
A. Pre-Incident Planning Checklist 351
B. Sample SOP/SOG for Engine Company Operations 354
C. NIOSH: Death in the Line of Duty Report 366
D. Tactical Work Sheets 383

Glossary 385
Index 393

Table of Contents

Preface viii
Introduction 1
Scope and Purpose 2
Book Organization 2
Key Information 3

1 Managing the Incident 9
National Incident Management System-Incident Command System 10
  Common Terminology 12
    Organizational Levels 12
    Resources 14
    Leadership Titles 14
  Modular and Scalable Organization 15
  Common Communications 15
  Unified Command Structure 16
  Unity of Command 17
  Incident Action Plan 17
  Manageable Span of Control 18
  Incident Facilities 18
  Comprehensive Resource Management 20
  Personnel Accountability 20
Incident Priorities, Strategies, Tactics, and Tasks 21
  Incident Priorities 22
  Strategies 23
  Tactics 23
  Tasks 24
Decision-Making 24
  Identify the Problem 25
  Determine a Solution 25
  Implement the Solution 26
Monitor the Results 26
  Results of Indecision 26
Understanding Your World and Local Resources 26
  Your World 27
  Your Resources 29
    Automatic Aid 32
    Mutual Aid 33
    Outside Aid 33
  Yourself 33
    Strengths 33
    Weaknesses 35

Preincident Planning 35
Summary 38
Review Questions 39

2 Fire Behavior, Building Construction, and Building Occupancy 43
Fire Behavior 44
  Fire Spread 45
  Fire Behavior in Compartments 47
  Rapid Fire Development 48
    Indicators 49
    Flashover 49
    Backdraft 51
    Rollover 52
    Smoke Explosion 53
    Autoignition 53
  Factors That Affect Fire Development 54
    Fuel Type 54
    Availability and Location of Additional Fuel 56
    Compartment Volume and Ceiling Height 57
    Ventilation 58
    Thermal Properties of the Compartment 59
    Ambient Conditions 59
    Fuel Load 59
Building Construction 60
  Construction Type 63
  United States Construction 63
    Type I (Fire Resistive) 64
    Type II (Noncombustible or Limited Combustible) 65
    Type III (Ordinary Construction) 65
    Type IV (Heavy Timber/Mill Construction) 66
    Type V (Wood Frame) 67
    Unclassified Construction Types 68
  Canadian Construction 69
  Building Periods or Age 70
    Early Regulatory Codes 70
    Suburban Development 71
    New Material Uses 71
  Interior Building Arrangement 72
    Open Floor Plans 72
    Compartmentalization 73

Basements, Cellar, and Crawl Spaces 73	Arrival Condition Indicators 131
Attics and Cocklofts 75	*Time of Day.. 131*
Concealed Spaces 76	*Weather 131*
Structural Collapse Potential 76	*Visual Indicators 132*

Occupancy Types **83**
    Single Use 83
    Multiple Use 83
Building Contents **88**
Summary **89**
Review Questions **89**

3   Size-Up: Evaluation and Assessment 93
Size-Up Application **93**
    Preincident 94
        Preincident Survey Process 97
        Required Fire Flow Calculations 100
        Before an Alarm 101
    While Responding 102
    Facts, Perceptions, and Projections 104
        Facts 104
        Perceptions.. 105
        Projections 105
    On Arrival 106
    During the Incident 111
Incident Size-Up Considerations **111**
    Facts 112
    Probabilities... 113
    Own Situation 114
    Decision-Making 114
    Plan of Operation 115
Critical Fireground Size-Up Factors **115**
    Building Characteristics 115
    Life Hazard 124
        General Considerations 124
        Residential 125
        Mercantile and Business 126
        Industrial 126
        Institutional 126
        Assembly 126
        Educational 126
        Unoccupied, Vacant, or Abandoned
          Structures... 127
    Resources 128
        Your Apparatus Resources 128
        Fire Protection Systems 128
        Water Supply 129
        Other Available Resources 130

Summary **136**
Review Questions **137**

4   Strategy 141
Incident Priorities **141**
    Life Safety 142
    Incident Stabilization 142
    Property Conservation 142
Risk vs. Benefit **143**
Operational Strategies **144**
    Offensive Strategy 145
    Defensive Strategy 145
Managing the Incident Scene **148**
    Develop Incident Action Plan 148
    Resource Tracking 149
    Implementing the IAP 152
        Command Options..................................... 153
        Resource Allocation 154
Lloyd Layman's Tactical Model **155**
Summary **156**
Review Questions **156**

5   Tactics 161
Rescue **161**
    Search Safety Guidelines 164
    Conducting a Search 165
        Primary Search... 166
        Secondary Search 167
        Victim Removal 167
Exposures **168**
    Interior Exposures 169
    Exterior Exposures 169
Confinement **171**
Extinguishment **173**
Overhaul **175**
Ventilation **178**
Salvage **181**
Summary **183**
Review Questions **183**

v

6 Residential Scenarios

Note to Instructors 188
Scenario 1 189
Scenario 2 199
Scenario 3 207
Scenario 4 215
Scenario 5 221
Scenario 6 229
Scenario 7 237
Scenario 8 243
Scenario 9 249

7 Commercial Scenarios

Note to Instructors 258
Scenario 1 259
Scenario 2 267
Scenario 3 273
Scenario 4 281
Scenario 5 287
Scenario 6 295

8 Special Hazards Scenarios

Note to Instructors 312
Scenario 1 313
Scenario 2 319
Scenario 3 327
Scenario 4 333
Scenario 5 341
Scenario 6 349

Appendix A
  Pre-Incident Planning Checklist 351

Appendix B
  Sample SOP/SOG for Engine Company
    Operations 354

Appendix C
  NIOSH: Death in the Line of Duty Report 366

Appendix D
  Tactical Work Sheets 383

Glossary 385

Index 393

List of Tables

Table 2.1	Comparison of Occupancy Classifications	84
Table 3.1	Estimating Water Supply Based On Tank Capacity	130
Table 5.1	Tactical Objectives and Intended Incident Priorities	161

Preface

The first edition of the IFSTA **Structural Fire Fighting: Initial Response Strategy and Tactics** manual is intended to provide fire personnel with the knowledge needed to deploy resources in the first 10 minutes of any structural fire incident. The manual is designed to help the reader develop a logical decision-making process for determining incident priorities, strategies, and tactics regardless of the available resources or configuration of the emergency services organization. Scenarios for residential, commercial, and special hazard incidents are included with recommended best practices and considerations. Training officers can use these scenarios, modified to local resources and realities, to develop classroom discussions or field simulations.

Acknowledgement and special thanks are extended to the members of the IFSTA validating committee. The following members contributed their time, wisdom, and knowledge to the development of this manual:

IFSTA Structural Fire Fighting: Initial Response Strategy and Tactics, First Edition
IFSTA Validation Committee

Committee Chair
William Neville
Neville Associates
Penn Valley, California

Committee Members

Albert Bassett
Norwalk Fire Department
Norwalk, Connecticut

Claude Beauchamp
Quebec National Fire Academy
Laval, Quebec, Canada

Brian Beresford
San Pasqual Reservation Fire Department
Valley Center, California

Paul Boecker III
Sugar Grove Fire Department
Oswego, Illinois

Bryan Collins
San Ramon Valley Fire Protection District
San Ramon, California

Vince Conrad
Halifax Regional Fire
Halifax, Nova Scotia, Canada

Kristofer DeMauro
City of Owasso Fire Department
Skiatook, Oklahoma

Duane M. Dodwell
Fairfax County Fire and Rescue Dept.
Waldorff, Maryland

Richard Dunn
Columbia Fire Department
Columbia, South Carolina

Jerry Hallbauer
Kansas Fire and Rescue Training Institute
Lawrence, Kansas

Larry Jenkins
Fairfax County Fire and Rescue Department
Alexandria, Virginia

Robert Kronenberger
Middletown Fire Department
Middletown, Connecticut

Committee Members (continued)

David Mager
Boston Fire Department
Franklin, MA

John McNeece
Maryland Fire and Rescue Institute
University of Maryland
College Park, Maryland

Matt Roberts
Phoenix Fire Department
Phoenix, Arizona

Demond Simmons
Oakland Fire Department
Oakland, California

Greg Willis
Wetumpka Fire Department
Wetumpka, Alabama

The following individuals contributed their assistance and comments as reviewers for this manual:

Ron (Raniero) L. Angelone
Indian River County Fire Rescue
Vero Beach, Florida

Jeff Jones
Tualatin Valley Fire and Rescue
Aloha, Oregon

Michael McLaughlin
Merced City Fire Department
Merced, CA

Harold Richardson
Yarmouth, Nova Scotia, Canada

Frank Roma
City of McKinney Fire Department
McKinney, Texas

The following individuals and organizations contributed information, photographs, and other assistance that made completion of this manual possible:

Wil Dane	Mike Mallory
Mathew Daly	McKinney (TX) Fire Department
Department of Defense	Chris Mickal
Department of Homeland Security	Ron Moore
Bob Esposito	Marvin Nauman, FEMA
Federal Emergency Management Agency (FEMA)	NIST
Dick Giles	Ed Prendergast
Iowa State Training Bureau	Tom Aurnhammer
Ron Jeffers	Tulsa Fire Department
Library of Congress	United States Fire Academy (USFA)

Additionally, gratitude is extended to the following members of the Fire Protection Publications **Structural Fire Fighting: Initial Response Strategy and Tactics Project Team** whose contributions made the final publication of this manual possible:

IFSTA Structural Fire Fighting: Initial Response Strategy and Tactics Project Team

Project Manager/Staff Liaison/Writer
Fred Stowell, Senior Editor

Editors
Libby Hieber, Senior Editor
Elkie Burnside, Student Intern

Technical Reviewer
Jeff Fortney, Senior Editor

Proofreader
Cindy Brakhage, Senior Editor

Photography
Jeff Fortney, Senior Editor

Production Coordinator
Ann Moffat

Illustrators and Layout Designers
Ruth Mudroch, Senior Graphic Designer

IFSTA Projects Coordinator
Ed Kirtley

Library Researcher
Susan F. Walker, Librarian

Editorial Assistant
Tara Gladden

The IFSTA Executive Board at the time of validation of the **Structural Fire Fighting: Initial Response Strategy and Tactics** manual was as follows:

IFSTA Executive Board

Chair
Jeffrey Morrissette
Commission on Fire Prevention and Control
Windsor Locks, Connecticut

Vice Chair
Paul Valentine
Mt. Prospect Fire Department
Mt. Prospect, Illinois

Executive Director
Mike Wieder
Fire Protection Publications
Stillwater, OK

Board Members

Stephen Ashbrock
Madeira & Indian Hill Fire Department
Cincinnati, Ohio

Steve Austin
CumberlandValley Vol. FF Assoc.
Newark, Delaware

Roxanne Bercik
Los Angeles (CA) Fire Department
Los Angeles, California

Mary Cameli
City of Mesa Fire Department
Mesa, Arizona

Bradd Clark
Owasso Fire Department
Owasso, Oklahoma

Dennis Compton
Mesa, Arizona

Frank Cotton
City of Memphis (TN) Fire Department
Memphis, Tennessee

George Dunkel
Scappoose, Oregon

John Hoglund
Maryland Fire and Rescue Institute
University of Maryland

Wes Kitchel
Santa Rosa (CA) Fire Department
Santa Rosa, California

Brett Lacey
Colorado Springs (CO) Fire Department
Colorado Springs, Colorado

Ernest Mitchell
Cerritos, California

Lori Moore-Merrell
International Ass. of Fire Fighters
Washington DC

Introduction

Introduction Contents

Scope and Purpose 2
Book Organization 2
Key Information 3

Introduction

Each year, the fire and emergency services in the United States respond to well over 20 million emergency responses. (There are, as of 2010, no similar statistics available for Canada.) Of these, approximately 500,000 involve fires in structures. Since the 1990s, the trend has been toward a decrease in structural fire responses and an increase in nonfire responses, primarily medical. For firefighters and fire officers, this trend means fewer opportunities to apply and practice their fire-suppression skills. For fire departments, this has meant greater reliance on live fire-training evolutions. The result has been an increase in firefighter injuries and a constant level of fatalities during both emergency incidents and training evolutions.

Data compiled by the National Fire Protection Association® (NFPA®) and the United States Fire Administration (USFA) indicate that the majority of all line-of-duty deaths (LODD) result from cardiac arrest. Many of these deaths have occurred during or shortly after emergency operations. The increased stress created by the emergency incident along with poor physical fitness, lack of annual medical evaluations, and poor nutrition have been leading causes in these deaths and heart-related injuries or illnesses. While each of these contributing factors can be controlled, they are beyond the scope of this manual.

In 2007, 115 firefighters died during emergency operations. The majority of the fatalities (54) were the result of cardiac arrest or stroke. Nineteen died by being lost, trapped, or incapacitated in a structure fire. The single deadliest incident resulted in nine firefighters being killed in a fast-moving fire in a retail furniture outlet. Noncardiac arrest fatalities can be reduced or eliminated by the application of strategy and tactics that allow the Incident Commander to have control over resources and their use during the incident.

Over 80,000 firefighters are injured at emergency incidents, with slightly less than half (48.6 percent) occurring during fire fighting activities. While the number of structure fires has decreased, injuries have increased by 5 percent per year. According to the NFPA®, the types of injuries that occur during fire fighting operations include the following:

- Strains, sprains, and muscular pain — 44.4 percent
- Wounds, cuts, bleeding, and bruises — 18.1 percent
- Burns — 7.0 percent
- Smoke or gas inhalation — 5.9 percent

It is a commonly accepted idea that fire fighting is a hazardous occupation that places personnel at risk throughout their work shift. It is also an accepted fact that many of the annual injuries and fatalities can be prevented through numerous means. This manual gives strategies and tactics that are intended to reduce casualties at structural fire incidents.

Although there have been numerous books on fire fighting strategy and tactics written over the past half century, the IFSTA **Structural Fire Fighting: Initial Response Strategy and Tactics** manual differs in some essential ways. First, unlike other current literature in the field, this manual has been reviewed and validated by a committee of fire and emergency service professionals of all ranks from across North America. Peer validation means that the committee reached a consensus agreement on the information contained in this manual.

Second, this manual takes a focused approach on strategy and tactics as they apply to the most frequent types of fires in occupancies: single- and multiple-family residential dwellings and small-to-medium-sized retail (mercantile) occupancies. In preparing to write this manual, it was determined that sufficient texts on high-rise, multi-alarm operations already exist. This new IFSTA manual provides the strategy and tactics needed to control average fires in average structures.

Finally, this manual is written for the initial response and the person, regardless of rank, who must evaluate the incident, assign resources, and attack the fire. The allocation of resources begins with the first-arriving unit and evolves to an average first assignment of apparatus, usually two engines and a ladder/truck company. The initial response ends with control of the incident, transfer of command to a battalion or district chief, or expansion of the incident management system to a higher level involving additional units or agencies.

Scope and Purpose

The scope of this manual is to provide the first arriving unit commander with the knowledge, skills, and abilities to assess the situation, initiate a command structure, and deploy resources until transfer of command or termination of the incident. The target audience is all career and volunteer personnel who will be assigned to incidents involving single and multifamily dwellings, commercial occupancies, and unique target or special hazards.

The purpose of the manual is to provide the initial IC with the strategic and tactical concepts that can be applied to various situations with the resources available to them. Nationally accepted strategies and tactics are presented and applied to hypothetical incidents throughout the manual. The manual is written for use in training agencies, college degree programs, and as a resource guide for all firefighters who are or may become unit commanders. The overall goal of this manual is to reduce fireground injuries and fatalities and increase the effectiveness of fire-suppression activities through better decision-making by the initial Incident Commander (IC).

Book Organization

This first edition of the IFSTA **Structural Fire Fighting: Initial Response Strategy and Tactics** manual is arranged in a manner designed to assist the reader in developing the knowledge, skills, and abilities required to manage an initial attack on a structure fire. Learning objectives and review questions help the reader focus on the topic of each chapter. Chapter titles and descriptions are as follows:

1. ***Managing the Incident*** — Provides an overview of the National Incident Management System-Incident Command System, introduces incident priorities, strategy, tactics, and tasks, discusses the decision making process, and provides information on local resources and the importance of preincident planning.

2. ***Fire Behavior, Building Construction, and Building Occupancy*** — Provides basic information on fire behavior, building construction, and occupancy and how they influence initial fire attack.

3. ***Size-Up: Evaluation and Assessment*** — Stresses the importance of an accurate size-up and how to achieve it.

4. ***Strategy*** — Describes incident priorities, risk management, and operational strategy, and introduces the theories of Lloyd Layman. Included are the Phoenix Model, Rules of Engagement, and incident action plan development and implementation.

5. ***Tactics*** — Details each element of the tactical model described in Lloyd Layman's theories.

6. ***Residential Scenarios*** — Contains sample scenarios for incidents involving typical residential occupancies. Scenarios are best practices based on information provided in Chapters 1 through 5.

7. ***Commercial Scenarios*** — Contains sample scenarios for incidents involving typical commercial occupancies. Scenarios are best practices based on information provided in Chapters 1 through 5.

8. ***Special Hazard Scenarios*** — Contains sample scenarios for incidents involving typical special hazard occupancies. Scenarios are best practices based on information provided in Chapters 1 through 5.

Key Information

Various types of information in this book are given in shaded boxes marked by symbols or icons. See the following examples:

> **Information Box**
> Information boxes give facts that are complete in themselves but belong with the text discussion. It is information that may need more emphasis or separation. They can be summaries of points, examples, calculations, scenarios, or lists of advantages/disadvantages.

A key term is designed to emphasize key concepts, technical terms, or ideas that firefighters need to know. They are listed at the beginning of each chapter and the definition is placed in the margin for easy reference.

Target Hazard — Facility or site in which there is a high potential for life or property loss.

Three key signal words are found in the text: **WARNING, CAUTION,** and **NOTE**. Definitions and examples of each are as follows:

- **WARNING** indicates information that could result in death or serious injury to fire and emergency services personnel. See the following example:

> **WARNING!**
> Live-fire training must adhere to the requirements set forth in NFPA® 1403, Standard on Live Fire Training Evolutions (2007).

- **CAUTION** indicates important information or data that fire and emergency services responders need to be aware of in order to perform their duties safely. See the following example:

> **CAUTION**
> Fire and emergency service responders must be familiar with the physiological, emotional, and technological limitations caused by the use of respiratory protection equipment to prevent injury or death.

- **NOTE** indicates important operational information that helps explain why a particular recommendation is given or describes optional methods for certain procedures. See the following example:

NOTE: This information is based on research performed by the International City/County Managers Association, Inc.

At the beginning of each chapter is a list of learning objectives to help focus attention on the information that the reader is expected to gain from the chapter. A chapter table of contents also provides an outline of significant topics and their page references. A list of key terms is included at the beginning of each chapter. Finally, at the end of the chapter, a series of review questions can be found to help ensure that you have read and learned the key points in the chapter. A glossary of important terms and appendices containing additional information conclude the manual.

Managing the Incident

Chapter Contents

National Incident Management System-Incident Command System 10
- Common Terminology 12
- Modular and Scalable Organization 15
- Common Communications 15
- Unified Command Structure 16
- Unity of Command 17
- Incident Action Plan 17
- Manageable Span of Control 18
- Incident Facilities 18
- Comprehensive Resource Management 20
- Personnel Accountability 20

Incident Priorities, Strategies, Tactics, and Tasks 21
- Incident Priorities 22
- Strategies 23
- Tactics 24
- Tasks 24

Decision-Making 24
- Identify the Problem 25
- Determine a Solution 25
- Implement the Solution 26
- Monitor the Results 26
- Results of Indecision 26

Understanding Your World and Local Resources 26
- Your World 27
- Your Resources 29
- Yourself 33

Preincident Planning 35
Summary 38
Review Questions 39

Key Terms

First Alarm Assignment 25	Target Hazard 36
Freelancing 26	Two-In/Two-Out Rule 35

Managing the Incident

Learning Objectives

After reading this chapter, students will be able to:

1. Discuss the characteristics of the National Incident Management System-Incident Command System.
2. Identify incident priorities.
3. Describe incident strategies, tactics, and tasks.
4. Explain each step of the decision-making process.
5. Describe your local response area and available resources.
6. Discuss preincident planning.

Chapter 1
Managing the Incident

As a fire officer, you are responsible for managing any emergency incident assigned to you. Your success as an incident scene manager will result from:

- Implementation of an Incident Command System (ICS)
- Selection of the best incident priorities, strategy, and tactics for the incident
- Good decision-making
- Efficient use of resources
- An awareness of your world including your resources, procedures, and personal strengths
- Application of knowledge gained from preincident surveys

The importance of a standardized Incident Management System has been recognized by most levels of government. The United States federal government even mandates the adoption of the National Incident Management System (NIMS) by all fire and emergency service organizations that accept federal funds. NIMS provides a model Incident Command System (ICS) structure that can be implemented for all types and sizes of emergency incidents.

In addition to the use of the ICS, Incident Commanders (IC) must also develop and adhere to a decision-making model that will result in the efficient and effective use of resources. That model must be simple enough for anyone to apply in a crisis and yet detailed enough to apply to all potential situations. Repeated use of a decision-making model is essential to successful allocation of resources and incident management.

> The authority having jurisdiction (AHJ) is responsible for determining the qualifications for personnel who may establish Incident Command regardless of that person's rank. In the remainder of this manual, the term IC or Command will be used to describe this person.

Fire officers and firefighters must also understand the basic strategies and tactics that will control, confine, and extinguish structure fires. These strategies and tactics are based on accepted models that have evolved over the years and on scientific research and experience.

Traditionally, the fire and emergency services have depended on experience to provide the basis for strategic and tactical decision-making. Responding to structure fires provided an opportunity for fire offficers and firefighters to gain valuable experience in decision-making as well as fire behavior. Over the

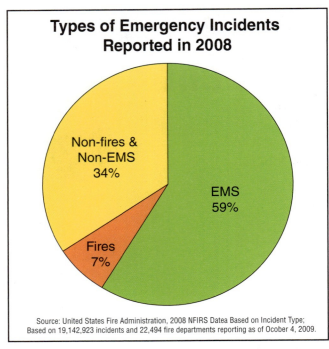

Figure 1.1 Comparison of types of emergency incidents in the U.S. in 2008. *Courtesy of the United States Fire Administration.*

past thirty years, however, improvements in life safety education, fire and building code enforcement, and the installation of fire alarm and protection systems has decreased the number of fires. The reduction in structure fires has decreased the opportunity to gain this level of experience. By 2008, the percentage of emergency responses involving fires was less than medical responses **(Figure 1.1)**.

Fewer structure fires means that ICs have fewer opportunities to apply a decision-making model during emergency responses and must rely heavily on information gathered in a classroom. Therefore, new fire officers and firefighters may lack the personal experience to make accurate decisions or apply the correct strategy and tactics to structure fires. This lack of actual experience must be counteracted with training, exercises, and simulations.

This chapter provides the basic information you will need to manage a structural fire incident. First, a basic ICS model based on the NIMS requirements is explained as it applies to the initial response. Second, a decision-making model is provided that is easy to learn and apply. Third, this chapter provides you with a strategic and tactical model for use with residential and commercial occupancies found in most response areas. Fourth, an awareness of local concerns is discussed including your awareness of your local resources and your personal strengths and weaknesses. Finally, you will see an overview of the preincident planning process that will provide you with information on target hazards and specific structures in your response area.

Taken together, the information provided here will give you, as the initial IC, the information you need to successfully manage structure fires in your jurisdiction. By building on successes in managing smaller, so-called routine incidents, you will be prepared to handle larger and more complex emergencies.

The resources discussed here are found in volunteer, combination, and career departments in initial response assignments. This manual references resources that are available from the first arriving unit to the transfer of command or incident termination. Regardless of department type or size, you must recognize that additional resources may be unavailable, limited, or delayed. These resource limitations will cause you to make decisions to efficiently allocate personnel and equipment.

National Incident Management System-Incident Command System

The basis for safe and efficient incident scene management is the Incident Command System (ICS). Various forms of ICS (sometimes called the Incident Management System [IMS]) have been developed, adopted, taught, and implemented by fire and emergency services organizations for many years. The creation of the National Incident Management System (NIMS) has generated an emphasis on the adoption and use of a standardized ICS model. Because the concept of incident management is so critical, the U.S. Government mandated the use of ICS as a component of NIMS in 2004.

This section provides a general introduction to the basic ICS model. A complete discussion of the NIMS-ICS model is provided in the National Incident Management System Consortium, *Book 1, Incident Command System Model Procedures Guide for Incidents Involving Structural Fire Fighting,* published by Fire Protection Publications.

The first-arriving IC implements the ICS, makes decisions, and takes actions that will influence the remainder of the incident. If resources are limited, the initial IC may be involved in rescue or fire suppression while making command decisions **(Figure 1.2)**. Whether the initial IC maintains command of the incident or transfers command to a senior or more experienced officer, the initial decisions must be reliable and based on the organization's ICS procedures.

By controlling emergency response resources (personnel, apparatus, equipment, and materials), the IC can effectively direct the actions that will stabilize the incident. ICS establishes an organizational structure for the control of these resources for all types and sizes of emergency incidents.

Every member of the organization must be familiar with the ICS system and trained in its application. In addition, all agencies that have mutual aid or automatic aid agreements in place must know and use the same system. Achieving this familiarity may require extensive cross-training at all organizational levels among units of the participating agencies, especially at the company level.

You must be familiar with the characteristics of the ICS that has been adopted by your emergency services organization. Common characteristics of the ICS include:

- Common terminology
- Modular and scalable organization

Figure 1.2 Limited resources may require the initial IC to be involved in fire suppression as well as command activities. *Courtesy of Bob Esposito.*

- Common communications
- Unity of Command
- Unified Command Structure
- Incident Action Plan (IAP)
- Manageable span of control
- Personnel accountability

Common Terminology

Emergency scene communication depends on common terminology for functional elements, position titles, facilities, and resources. The following terms describe organizational levels, resources, and leadership titles that all fire and emergency service responders need to understand.

Organizational Levels

The ICS is described with terms that define the basic levels of organization, responsibility, and authority. Those terms include:

- *Command* — Act of directing, ordering, and/or controlling resources by virtue of explicit legal, agency, or delegated authority; also used to denote the organizational level that is in overall command of the incident. Lines of authority must be clear to all involved. A clear chain of command must be followed with each person reporting to one supervisor and following that supervisor's orders.

- *Command Staff* — Assigned on larger incidents, or as required, to perform key functions of the IC, but are not part of the line organization. The Command Staff works for and advises the IC and includes the public information officer, safety officer, and liaison officer. On small incidents, the IC will perform most or all of these duties. The position of Incident Safety Officer (ISO) may be delegated as the IC sees fit **(Figure 1.3)**.

Figure 1.3 The Incident Safety Officer is an essential member of the Command Staff.

- *General Staff* — Incident management personnel who represent the major functional Sections. General Staff positions include Operations, Planning, Logistics, and Finance/Administration. Like the Command Staff, the initial IC will perform some of these duties as needed until it is appropriate to delegate them to others as the incident expands.

- *Section* — Organizational level located between Command and Branch having responsibility for a major functional area of incident management. Establishing Sections occurs as the ICS is expanded to respond to the increase in the size or complexity of the incident.

- *Branch* — Organizational level having functional and/or geographic responsibility for major segments of incident operations; organizationally located between Section and Division or Group. Branches are identified by roman numeral or functional area (such as Fire, Multi-Casualty, Law, Support, Service, etc).

- *Division* — Operational level having responsibility for tactics within a defined geographic area; organizationally between Branch and Single Resources, Task Force, or Strike Team. Resources assigned to a division report to that division leader. Examples:
  — Divisions are assigned clockwise around an outdoor incident with Division A at the front of the incident.
  — In multistory buildings, Divisions are usually identified by the floor or area to which they are assigned: first floor is Division 1, second floor is Division 2, etc.

- *Group* — Organizational level equal to Division, having responsibility for a specified functional assignment at an incident (such as ventilation, salvage, water supply, etc.) without regard for a specific geographical area **(Figure 1.4)**. Resources assigned to a Division report to that Group Leader.

- *Unit* — Organizational level within the Sections that fulfill specific support functions such as the resources, documentation, and demobilization within the Planning Section.

Figure 1.4 Operating under the direction of a Group Leader, the Ventilation Group coordinates its activities with other Groups.

Resources

The following terms are used to describe and define the resources that the IC has available or potentially available for assignment. Resource designations provide an indication of the type and capability of the resources. It is important to track the status of these resources so that they may be assigned without delay when and where they are needed.

- ***Resources*** — All personnel and major items of equipment, supplies, and facilities; may include individual companies (single resources), task forces, strike teams, or other specialized units. Factors:
  - Resources are considered available when they have checked in at the incident and are not currently committed to an assignment.
  - Resources may be located in the Staging Area.
- ***Crew*** — Specified number of personnel assembled for an assignment such as search, ventilation, or hoseline deployment and operations. The number of personnel assigned to a crew should be within span-of-control guidelines.
- ***Single Resources*** — Individual pieces of apparatus (engines, ladders/trucks, water tenders, bulldozers, air tankers, helicopters, etc.) and the personnel required to make them functional.
- ***Task Force*** — Any combination of resources (engines, ladders/trucks, bulldozers, etc.) assembled for a specific mission or operational assignment. All units in the force must have common communications capability and a designated leader.
- ***Strike Team*** — Set number of resources of the same kind and type (engines, ladders/trucks, bulldozers, etc.) that have an established minimum number of personnel as established by the AHJ. All units in the team must have common communications capabilities and a leader in a separate vehicle.

Leadership Titles

Finally, consistent terms are used to describe leadership roles within the ICS. They include:

- ***Chief*** — Title of individuals who are assigned as a member of the general staff and are responsible for the Sections.
- ***Director*** — Title of individuals who are responsible for the management of a Branch. Branches typically exist in the Operations and Logistics Sections; however, a Branch can be established, as needed, in any Section to maintain the span of control.
- ***Incident Commander (IC)*** — Title of the individual who is responsible for the management of all aspects of the incident. Primarily responsible for determining the incident priorities, formulating the Incident Action Plan (IAP), and coordinating and directing all incident resources to implement the plan and meet its objectives **(Figure 1.5)**.
- ***Leader*** — Title of individuals who are responsible for the management of a Unit, Strike Team, or Task Force. Strike Teams and Task Forces are operational resources, while Units fulfill support roles.
- ***Manager*** — Title of individuals who are responsible for an area or facility. This title applies to the staging area, base/camp, and aircraft facilities.

- *Officer* — Title of individuals who are members of the Command Staff and responsible for safety, liaison, and public information.
- *Supervisor* — Title of individuals responsible for command of a Division or Group within the Operations Section; may be assigned to an area initially to evaluate and report conditions, and advise Command of the needed task and resources.

Modular and Scalable Organization

The ICS is intended to be modular and scalable. Modular design means that the incident command structure is based on functional modules **(Figure 1.6a)**. Each module has assigned tasks and resources to fulfill them. Based on the type, size, and complexity of the incident, the modules can be activated and added to the command structure. Unneeded modules are left vacant and modules that complete their tasks as the incident nears termination can be deactivated.

Figure 1.5 The IC must remain in communication with all units at the incident, units en route, and the communication center.

Scalable design means that the command structure expands as needed from the top down depending on the complexity of the incident **(Figure 1.6b, p. 16)**. Initially, the IC will perform all of the command duties required by the incident. However, as the incident becomes more complex and additional resources arrive, the IC delegates the command duties to other personnel. The initial IC remains in charge of the incident until command is transferred to a relieving officer or the incident is terminated.

Common Communications

A common means of communication is essential to maintaining control, coordination, and safety at any incident. ICS requires that a common communication system provides the ability to be understood and to contact all units or agencies that are assigned to the emergency incident. To ensure ef-

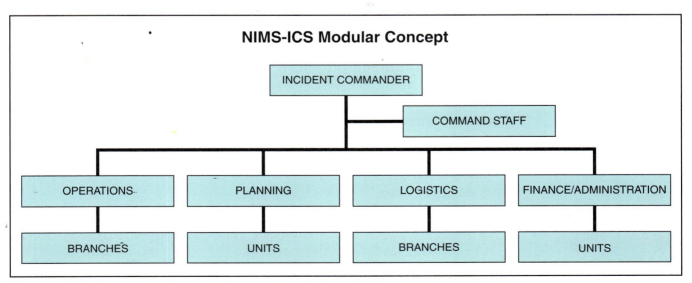

Figure 1.6a The modular concept.

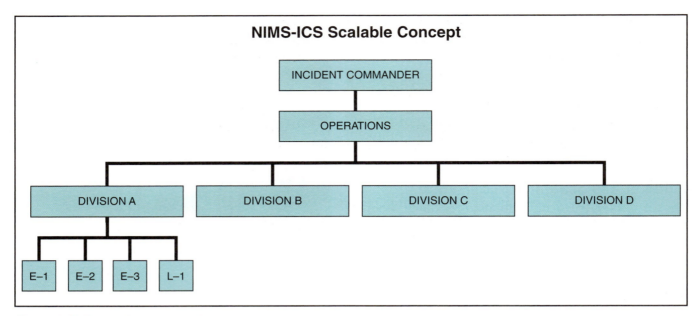

Figure 1.6b The scalable concept.

fective communications, all units must use clear text (specified phrases in plain English) rather than the numeric 10-codes, other agency-specific radio codes, or jargon.

The ICS also requires the establishment of a common incident communications plan that identifies different radio frequencies or channels to be used exclusively for specified organizational functions. To avoid the chaos that would result from all units attempting to receive and transmit on the same channel, the Incident Communications Plan assigns specific channels to specific functions or units.

While most modern mobile and portable radios are capable of scanning, receiving, and transmitting over dozens of channels, not every organization is equipped with the latest communications equipment. If mutual aid units are not equipped with radios that can receive and transmit on the channels assigned to them in the plan, they must be issued portable radios that will function on those channels.

Unified Command Structure

A Unified Command Structure is necessary when an incident involves or threatens to involve more than one discipline (i.e., law enforcement, public works), jurisdiction, or agency. For example, a fire that originates in a wildland/urban interface can rapidly spread across jurisdictional boundaries. However, these multi-jurisdictional incidents are not limited to fires. Natural disasters frequently encompass multiple communities and levels of governmental authority.

A Unified Command may also be appropriate within a single jurisdiction if multiple disciplines or agencies are affected. For example, a fire in a clandestine drug laboratory may begin as a fire department incident, but then require assistance from the local law enforcement agency or state and federal authorities when the true nature of the operation is recognized.

In a Unified Command Structure, representatives of all affected agencies or jurisdictions share the Command responsibilities and decisions. They jointly establish one set of objectives and strategies for the incident and agree on the tactics that must be performed. Regardless of the nature or magnitude of the incident, only one Operations Section is established per incident. In some agencies or jurisdictions, legal authority to act is vested in those occupying certain positions of responsibility. Unified Command allows these individuals to interface with those who have the operational expertise required to resolve an incident.

As the fire and emergency service becomes more involved in all-hazard types of responses, it is essential that personnel become more aware of the Unified Command process. Even small incidents may result in a situation that requires a Unified Command approach. Cross-jurisdictional incidents will also place the company officer either in charge or as part of a Unified Command.

Unity of Command

The concept of Unity of Command is essential to ensure the accountability and effectiveness of personnel operating within the ICS. Unity of Command means that each resource assigned to the incident reports to only one supervisor. At a small incident, the initial IC supervises his or her own crew and the officers of each additional unit while those officers supervise their own crews **(Figure 1.7)**.

Figure 1.7 A single company officer supervising an attack hose team is an example of Unity of Command. *Courtesy of Bob Esposito.*

Incident Action Plan

NFPA® 1021, *Standard for Fire Officer Professional Qualifications,* requires the Fire Officer I to be able to develop an Initial Action Plan. The NFPA® term *Initial Action Plan* means the same thing as the ICS Incident Action Plan (IAP); that is, a written or unwritten plan for the safe and efficient resolution of an emergency incident. Chief Lloyd Layman and other authors have written extensively about the size-up process, and Layman uses the term *plan of operation* to describe the same concept. According to NFPA® 1561, *Standard on Emergency Services Incident Management System*, an IAP establishes the overall strategic decisions and assigned tactical objectives for an incident.

Even at a small incident that does not require a written IAP, the IC must create an action plan and communicate it to those who will implement it. An IAP is an integral part of ICS, and implementing the plan affects how emergency resources are organized. All personnel assigned to an incident must function according to the IAP where incident priorities, strategies, tactics, and support requirements for the incident are formulated. The initial IC takes the first steps in creating an IAP that can evolve as the incident expands to involve more resources.

NOTE: Tactical work sheets can be used at incidents that do not require a written IAP. These work sheets can become the basis for the written IAP if the incidents become more complex.

At a small incident, the initial IC determines the incident priorities, selects the overall strategy for dealing with the incident, and establishes the tactics for meeting that strategy. The IAP contains all tactical and support activities required for the control of the incident **(Figure 1.8)**. The plan is divided into operational periods consisting of specific time intervals. The duration of the operational periods may vary, depending on the complexity and type of incident, the estimated time to terminate the incident, the number of units and agencies involved, and environmental and safety considerations. Generally, operational periods may be as short as 2 hours or as long as 24 hours with the average being 12 hours.

Manageable Span of Control

The principle of span of control defines the number of direct subordinates that one supervisor can effectively manage. Variables such as proximity between the supervisor and subordinates, similarity of function, complexity of the incident, and subordinate capability affect that actual number. An effective span of control ranges from one to seven subordinates per supervisor, depending upon the variables mentioned above, with five considered to be the optimum number **(Figure 1.9, p. 20)**. When an effective span of control is maintained, it is much easier for supervisors to keep track of their subordinates and monitor their safety and their activities. Situations that permit an increase in the number of subordinates one supervisor can control include situations when:

- Subordinates are within sight of the supervisor and are able to communicate effectively with each other
- Subordinates are performing the same or similar functions
- Subordinates are skilled in performing the assigned task

Incident Facilities

According to NIMS, several possible types of facilities can be established in and around an incident. The titles and functions of these facilities have been designated by the NIMS-ICS model. The types of facilities and their locations are determined by the requirements of the incident as outlined in the IAP. The most commonly used incident facilities for small incidents are:

INCIDENT OBJECTIVES	1. INCIDENT NAME Billings Warehouse	2. DATE PREPARED 1/25/2007	3. TIME PREPARED 0925

4. OPERATIONAL PERIOD (DATE/TIME)
1/25 0915 to Completion

5. GENERAL CONTROL OBJECTIVES FOR THE INCIDENT (INCLUDE ALTERNATIVES)
Contain fire to northwest corner of warehouse area
Use building fire suppression system to control and eliminate fire
Remove smoke and fire gases from structure to minimize damage to structure
Remove water from fire area
Conduct search of fire area when safe

6. WEATHER FORECAST FOR OPERATIONAL PERIOD
Clear, 35 F, Wind out of southwest at 5 mph, humidity 10%

7. GENERAL SAFETY MESSAGE
Be aware of potential roof collapse hazard
Respiratory protection & pass devices required in all operational areas.
ISO's assigned to all branches
Rehab located at corner of B-C

8. ATTACHMENTS (🕒 IF ATTACHED)
- ☐ ORGANIZATION LIST (ICS 203)   ☐ ORGANIZATION LIST (ICS 203)   ☐ _____
- ☐ ASSIGNMENT LIST (ICS 204)   ☐ ASSIGNMENT LIST (ICS 204)   ☐ _____
- ☐ COMMUNICATIONS PLAN (ICS 205)   ☐ COMMUNICATIONS PLAN (ICS 205)   ☐ _____

202 ICS 3-80	9. PREPARED BY (PLANNING SECTION CHIEF) J. E. FORTNEY	10. APPROVED BY (INCIDENT COMMANDER) E. C. KIRTLEY

Figure 1.8 Sample Incident Action Plan.

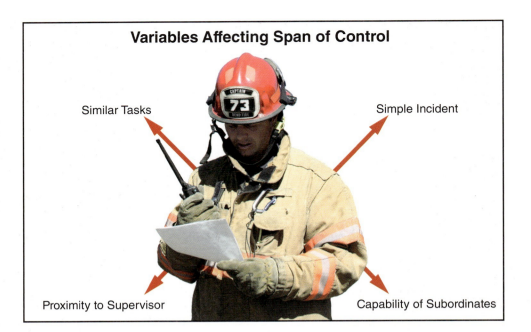

Figure 1.9 Variables that affect the span of control of a supervisor.

- ***Incident Command Post (ICP)*** — Location from where all incident tactical operations are directed; located at or in the immediate vicinity of the incident site
- ***Staging areas*** — Any location where resources (personnel, supplies, and equipment) are held in reserve while awaiting operational assignment

Additional facilities used at expanded incidents are discussed in the IMS Consortium manual mentioned earlier.

Comprehensive Resource Management

The purpose of comprehensive resource management is to provide the IC with access to and control over all available resources. Resource management involves maintaining an accurate and up-to-date picture of resources for utilization to include personnel, teams, equipment, supplies, and facilities.

Personnel Accountability

An essential element of the ICS is personnel accountability, which ensures the efficient use of resources and the safety of all those involved in an incident. Supervisors are responsible for knowing where their subordinates are at all times and what tasks have been assigned to them. Therefore, the initial IC must implement the personnel accountability system for crew members, additional units, and individuals reporting to the incident.

Several different types of accountability systems are presently in use (**Figures 1.10a&b**). Emergency incident accountability systems are usually dictated by the organization's policies. Procedures for using an accountability system must be followed closely by all personnel. The IC is accountable for each individual and unit assigned to the incident.

The NIMS-ICS provides a means of tracking the personnel resources assigned to a given incident. Personnel accountability includes all the following elements:

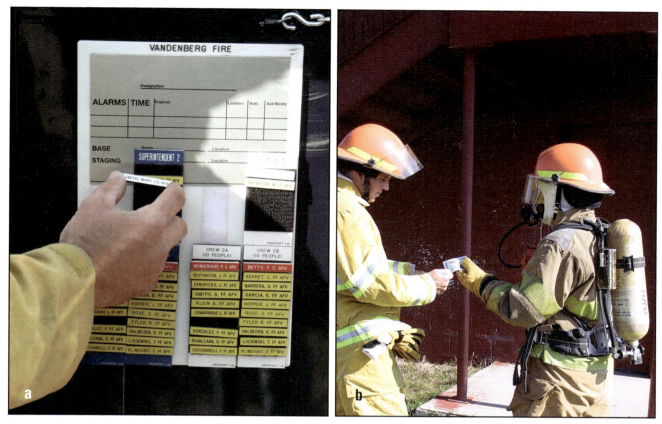

Figures 1.10a&b These photographs depict two of many types of accountability systems that are available.

- ***Check-in*** — Requires all responders, regardless of affiliation, to check in with the IC to receive their assignments
- ***Incident Action Plan (IAP)*** — Identifies incident priorities and objectives, which dictate how strategic and tactical operations must be conducted
- ***Unity of Command*** — Dictates that each responder has only one supervisor
- ***Span of Control*** — Gives supervisors a manageable number of subordinates
- ***Resource tracking*** — Ensures that each company officer reports resource status changes as they occur

When personnel at all levels in the emergency services organization operate according to these principles and procedures, personnel accountability and safety are maximized. As an incident grows from an initial alarm assignment to a major incident, these basic principles and procedures must continue to be applied.

Incident Priorities, Strategies, Tactics, and Tasks

To fully understand incident priorities, strategies, tactics, and tasks, their key components must first be defined. As you read this manual, you should think of incident priorities, strategies, tactics, and tasks as a pyramid or hierarchy of concepts. At the top is the incident priority or objective which is the desired outcome of the incident. Next is the choice of strategies needed to fulfill that

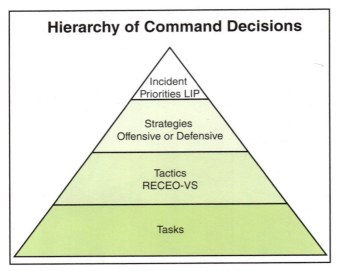

Figure 1.11 Hierarchy of Command Decisions

priority. The desired strategy is accomplished by the completion of certain tactics. Tactics are accomplished by performing tasks which may require the efforts of one or more individuals or units, who actually do the work **(Figure 1.11)**.

Incident Priorities

Incident priorities, sometimes referred to as incident objectives, are determined by the IC and are basic to the entire incident. They are best described by the acronym LIP, which stand for:

- Life safety
- Incident stabilization
- Property conservation

Life safety involves the saving or protecting of all human life at the incident beginning with yourself and the personnel assigned to you. You must evaluate the level of risk you expose your personnel to in an attempt to save any victims of the incident. You must never place yourself or crew in jeopardy in an attempt to save a life that is already lost. In the same way, you must never place your personnel at risk to save property that is already lost **(Figure 1.12)**.

Incident stabilization means controlling the fire and eventually extinguishment. This priority may be implemented in stages, first by locating the fire, then by confining it to the compartment or structure of origin, and then by directing extinguishing agent onto the fire. The overhaul tactic will complete this priority as hidden or remaining fires are extinguished.

Finally, property conservation is the last incident priority. Unfortunately, many firefighters have been killed or injured attempting to save property that was not savable. Remember, physical property can be replaced, human life cannot. See Chapter 4 for additional information on incident priorities.

Figure 1.12 Rapid fire development or a delayed response may result in a fire that is completely beyond control. *Courtesy of Chief Chris Mickal, New Orleans (LA) Fire Department.*

LIP is intended to help you begin the decision-making process when you arrive at the incident. From that point, you can determine the strategy that will best mitigate the incident.

Strategies

There are two strategies from which you can select from to meet your incident priorities. These strategies are offensive and defensive. An offensive strategy consists of an interior fire attack intended to stop the fire in the compartment or area of origin and prevent it from spreading to uninvolved areas **(Figure 1.13a)**. A defensive strategy is used to confine the fire to the structure of origin and prevent it from spreading to exterior exposures (adjacent structures) **(Figure 1.13b)**. See Chapter 4 for additional information on strategies.

Tactics

Tactics are activities used to achieve an immediate or short-term objective that, when completed, fulfill the incident priorities. They are performed at the division, group, and unit level of the ICS. Supervisors direct their resources in operational activities (tasks) to accomplish tactical objectives. A tactical level assignment comes with the authority to make decisions and assignments within the boundaries of the overall plan and safety conditions. Examples of tactics are performing a primary search of an area, confining a fire to a structure or compartment, or using vertical ventilation to remove smoke and heated gases from a structure.

Tactical objectives must be measurable, attainable, and flexible. Measurable objectives have specific outcomes that are obvious to the observer, such as the opening of a door through forced entry. An attainable objective is one that can be accomplished by the resources assigned to it, in this case, two firefighters who have the skills and tools to force open a door. Finally, the tactical objective must be flexible based on rapid changes in the situation. For instance, if the tactic is to remove smoke from the structure through vertical ventilation, the weakening of the roof may require a shift to forced ventilation through a window or door.

Figures 1.13a & b (a) The use of small handlines and an interior attack typify an offensive strategy while (b) the use of master stream appliances, large-sized hose lines, and an exterior attack typify a defensive strategy.

Tasks

Tasks are activities normally accomplished by individuals, companies, or crews. Tasks are routinely supervised by company officers or senior firefighters. The accumulated achievements of tasks should accomplish the desired tactics **(Figure 1.14)**. When tasks are completed, the IC must be informed of the completion. An example of a task is placing ground ladders against a building to gain access to the roof and cutting a hole for vertical ventilation. The placing of the ladder and cutting the hole are tasks that contribute to the ventilation tactic.

Decision-Making

Developing an IAP, determining the incident priorities, strategies, and tactics, and assigning resources require a sound decision-making process. During an emergency incident, the IC is under pressure to make the right decisions quickly and having a preexisting model greatly aids in that process.

All emergency scene decision-making should be based on a sound model. Therefore, a decision-making model contains steps that are taken in sequence to reach the desired outcome. The steps of the decision-making process should be simple so that they can be carried out rapidly under pressure. However, the process must lead to an action. Without the action step, the process is only a good intention.

By applying the model in daily activities, you will become more confident and able to apply it in an emergency. The steps of the process are:

- Identify the problem.
- Define the best solution.
- Implement the solution.
- Monitor the results.

As the situation changes, it may be necessary to go through the cycle again by identifying any new problems that have resulted from implementation of the solution. Decision-making is a dynamic process that continues until the incident is terminated **(Figure 1.15)**.

Figure 1.14 Deploying a master stream appliance is an example of a task that firefighters perform to complete a tactical objective.

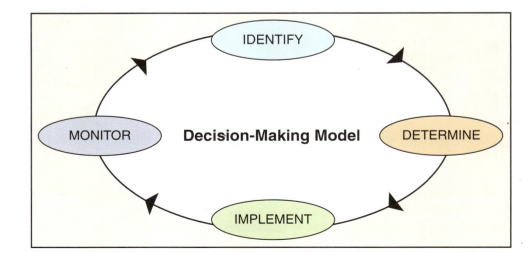

Figure 1.15 Decision-making is a dynamic process symbolized by a cycle or loop.

Identify the Problem

To identify the problem, you must first gather facts based on the dispatch report, the preincident survey, your personal observations, and your experience. Once gathered, the facts must be analyzed. The analysis will identify and define the problem.

For example, your company has been dispatched to a report of smoke in a single-story single-family dwelling. The time is 3:00 am on a Monday morning. Your company is fully staffed with 4 people and a complete **first alarm assignment** has been dispatched. According to the dispatch report, the occupants have evacuated the structure and reported a fire in the master bedroom. Upon arrival, you identify the problem as a working fire. At this point, you determine if the incident priority is life safety, incident stabilization, or property conservation. Keep in mind that the example in this and the following sections addresses only one problem. You must recognize that multiple problems may exist requiring multiple solutions **(Figures 1.16a - b)**.

> **First Alarm Assignment** — Initial fire department response to a report of an emergency; the assignment is determined by the local authority based on available resources, the type of occupancy, and the hazard to life and property.

Figures 1.16a&b (a) In some jurisdictions, a first-alarm assignment may consist of a pumper and water tender (b) in other jurisdictions it may include pumpers, aerial devices, and support apparatus. *Photograph courtesy of Oklahoma City (OK) Fire Department.*

Determine a Solution

When the problem has been identified, you need to determine a solution. If resources and time permit, alternate solutions should be determined and the best solution chosen. You may have to establish limits on the alternate solutions based on knowledge of local standard operating procedures (SOPs), available resources, or time available to implement the solution. In most instances, alternative solutions may be limitless.

The best solution is the one that results in the best outcome. The following questions should be considered when making your decision:

- Has the safety of everyone, including firefighters, been considered?
- Are there enough resources to correct the problem?
- What will happen if I don't do anything?

In the example given above, your first action is to establish command and give a brief initial radio report to the dispatch center and the other units that are part of the assignment. Based on the information that you have, you are reasonably certain that occupant lives are not in danger and rescue is not needed. Your primary incident priority becomes incident stabilization which will also accomplish property conservation through an offensive strategy. Your best solution is to apply water to extinguish the fire in the bedroom while ventilating the structure.

Implement the Solution

Your resources will implement the solution by performing tasks such as establishing horizontal or vertical ventilation, setting ladders, and advancing hoselines. In fire fighting terms, the decision becomes an action when the tactics are accomplished by performing the necessary tasks. In our example, an interior attack crew advances a hoseline into the structure while additional personnel perform ventilation.

Monitor the Results

You determine whether the decision has effectively solved the original problem through feedback from radio reports and face to face communication from other officers or firefighters and personal observation. Feedback and observation will also tell you whether other problems have been created by the initial decision. In the example, you are monitoring the effectiveness of the hose stream in controlling the fire and of ventilation through the removal of smoke from the compartment or structure.

Result of Indecision

While some types of problems can be resolved by making no decision, indecision is not an option at an emergency incident. The lack of action can and will result in greater damage, a potential for loss of life, and/or rapid fire spread. An effective IC must make sound decisions based on experience and training to mitigate the situation.

Indecision on the part of an IC can result in **freelancing**. Freelancing occurs when individuals or crews decide to take action they believe will help to control the situation. Such actions effectively remove them from the command and control of the IC. Their actions may even increase the hazard to other personnel and to themselves and decrease the effectiveness of the actions ordered by the IC.

> **Freelancing** — Operating independently of the IC's command and control.

Understanding Your World and Local Resources

Knowledge of the world in which you operate is critical to successfully controlling an emergency incident. Your world is comprised of your response area, your resources, and yourself. Your world will be different from other response areas whether they are across town or across the country. You must be aware of the characteristics of your world and have accurate knowledge of the resources at your disposal. You must also be aware of your own personal strengths and weaknesses. Your strengths may include, among others, your training, leadership and communication skills, and situational awareness. You may have weaknesses in any of these areas, which you will have to recognize and overcome through training.

Your World

Your world consists of your primary response area. Sometimes referred to as your first due or first in area, it may be as small as a few square miles of densely populated urban area to more than a 100 square miles (258 k²) of rural agricultural land or forest **(Figures 1.17a - b)**. Within that area, multiple types of occupancies, ages of structures and residents, types of building construction, and barriers to rapid response may exist. With each day, changes will occur within your response area that will alter existing hazards you will face or how you will respond to those hazards.

Because life safety is your first priority, you must be familiar with the population demographics within your response area. This includes the age, population density, and behavioral characteristics of the individuals and groups that populate the area. You should know the types and locations of high life hazards and become familiar with those places through preincident building surveys and preplans.

The structures that people live and work in should be your second concern. Information about the structures that you need to consider include:

- Types of building construction
- Types of building materials used
- Age of construction
- Application or lack of building code requirements at the time of construction or alteration
- Zoning requirements that determine the type of occupancies in a given area
- Separation distances between structures
- Presence or lack of fire protection systems
- Modifications/alterations made to existing structures
- Changes in building and fire code requirements

Figures 1.17a&b Across the country, response areas include (a) urban areas that are densely populated and filled with high rise buildings and (b) rural areas that are sparsely populated with two-story buildings.

Your next concern is the condition and design of streets, roads, and highways that you must use to reach the incident site. You must be familiar with permanent characteristics such as:

- Location of controlled and uncontrolled intersections
- Topographical features such as blind hills, sharp curves, or dips in the road surface
- Limitations on apparatus speed such as school bus stops
- Restrictions on apparatus weight due to types of road surfaces
- Limitations created by private roadways
- Overhead obstructions such as wires, canopies, or toll gates
- Bridge weight restrictions

- Railroad crossings **(Figure 1.18)**
- Tunnels, underpasses

Temporary characteristics will confront you on a daily basis. They include:

- Construction both on and off the road surface **(Figure 1.19)**
- Traffic congestion during peak commute times
- Traffic congestion due to events or accidents
- Pedestrian traffic in entertainment districts
- Farm traffic during harvest times
- Weather conditions such as rain or snow

While the previous concerns relate to all types of emergency responses, water supply is a primary consideration relating to structure fires. You must be familiar with:

- Water distribution system including water main size and capacity
- Hydrant spacing and locations
- Available water flow pressure and capacity
- Peak times of water usage
- Location of static water supplies
- Diameter and capacity of apparatus supply hose
- Apparatus pump capacities

Figure 1.19 Road construction causes traffic congestion that will slow response times.

Figure 1.18 Railroad crossings may create a barrier that delays responses as well as creating a hazard to responders.

Figure 1.20 Occasionally, heavy flooding can make roadways impassible. *Courtesy of Marvin Nauman, FEMA Photographer.*

Finally, you must be able to recognize the effects that weather has on your ability to respond and to control a structure fire. Weather conditions that you must consider include:

- Wind direction and velocity
- Relative humidity
- Rain, snow, or ice
- Flooding (**Figure 1.20**)
- Temperature (heat index or wind chill)

Your Resources

The most basic resources at your disposal at the incident are the personnel, tools, equipment, water supply, and apparatus assigned to you. You should be very aware of their capabilities and limitations. While the apparatus and its contents may remain constant, your personnel may not.

The capabilities of your personnel will vary. You must take into consideration how variables, such as staffing levels, temporary transfers and replacement personnel, training, and experience, will affect the safe and efficient operations at the incident. As you will see, establishing an Incident Action Plan will help to determine the best use of personnel resources. Other variables, such as the mental, emotional, and physical condition of your subordinates and yourself will also vary and affect the efficiency and safety of the operations.

Your ability to control an emergency also depends on the condition and capabilities of the apparatus, tools, equipment, and water supply that you bring to the incident scene. For example, before making an offensive interior attack with the water available on your apparatus, you must consider how long it will be before you have a constant water supply from a water distribution system, a relay from a static source, or a water tender. You must also consider whether you have sufficient personnel to perform the required hoseline attack.

Regardless of your response area, you must be familiar with your local public water supply distribution system, the location of operating hydrants, and their ability to supply a sufficient quantity and pressure of water. This knowledge extends to static water supplies such as swimming pools, ponds, and lakes. While the first arriving unit may not have the time to establish a drafting operation, you must realize that other units will require time and personnel to do so **(Figure 1.21)**. Where private water supply systems exist, you must be aware of its capacity and pressure as well as the location and operation of control valves. Establishing a primary and secondary water supply will normally guarantee sufficient water to extinguish most fires.

Figure 1.21 Establishing a drafting operation can be both time and labor intensive.

Caution
Always consider the potential need for a secondary water supply.

How you communicate with the dispatch or communication center and other responding units is also an important element of your local resources. Two-way radio communication systems are currently the standard for most fire and emergency services. You must know the operating protocol for these systems as well as their limitations due to topography and structural interference. In some cases, backup systems based on cellular telephones or landlines may be necessary to communicate with the communications center. Volunteer organizations may also depend on pagers and cellular telephones to alert members. The effectiveness of these devices will have an effect on response times for each individual.

Time is of the essence in modern fire fighting. The term *reflex time* has been established to describe the time between receipt of an emergency call to the point that action is being taken by responders at the incident. The total reflex time sequence consists of five segments. However, there is also a pre-reflex time sequence that occurs between ignition, discovery, reporting, and dispatch. You have no control over these times and they will vary.

Total reflex time includes:

- **Dispatch time** — Dispatch time is the time required to receive the call and notify the proper response units and is controlled by the efficiency of the dispatch or communication center. The level of technology, number of call-takers or dispatchers, and the operational protocol can determine the time required to:
  - Receive the call
  - Determine the type of emergency
  - Verify the location of the emergency
  - Determine the type and number of resources required for the emergency
  - Notify the resources about the emergency

- **Turnout time** — Turnout time is the time between receipt of the dispatch notification to the beginning of the response. It can be controlled through station design and layout, apparatus design, and department policy and procedures. You can reduce turnout time through proper placement of personal protective equipment, familiarity with your duties and location on the apparatus, and training **(Figure 1.22)**.

- **Response time** — Response time begins when the units start toward the incident until they arrive on scene. Referred to as wheel start to wheel stop in some documents, response time is primarily controlled by station location within a specified response area. Dispatching protocol that selects the closest units when the first response units are out of service also provides the shortest response times. Thorough, accurate, and current knowledge of your response area including traffic congestion, construction, and other potential obstructions allows you to manage your response time.

- **Access time** — Access time is the time it takes for firefighters to move from the apparatus to the point where the emergency exists including moving to upper stories of a building or dealing with any physical barriers such as locked doors or gates. You can control access time through familiarity with the incident scene. For structures and target hazards, this familiarity is gained through frequent and thorough preincident surveys. Preincident plans created from the information gathered in these surveys should contain information about any potential barriers, the best access points, and water supply sources among other information.

- **Setup time** — Setup time is the time required to deploy and connect hoselines, set ladders, and prepare to control the emergency. This is sometimes described as the point when attack lines are discharging water onto a fire. Training is the key to reducing setup time. Each member of the company must be familiar with their assigned tasks, the location of equipment on the apparatus, and the Incident Action Plan as developed by the IC.

Figure 1.22 Practicing donning personal protective equipment prior to mounting the apparatus can reduce unit turnout times.

You can have some control over turnout, response, access, and setup times. Decreasing these times is the responsibility of all department members from administration through company officers and firefighters.

Finally, the location of resources in the service area will influence the reflex time for both the initial unit and subsequent units. In areas served by urban career departments, stations and units are located to maintain a reasonable response time. In rural areas, response times are usually higher, and in volunteer organizations the response times are determined by the arrival of individuals in privately owned vehicles (POVs) in addition to the apparatus **(Figure 1.23)**. As an IC, you must recognize that you may be the only person or unit on the scene for an undetermined period of time. How you use this time will determine how well the emergency is managed. It is often said the first five minutes of an incident will determine the next five hours.

Tied to response times are the resources supplied through interagency or governmental agreements. Agreements with other neighboring jurisdictions are generally created to provide additional resources, provide coverage in adjacent areas, or reduce the cost of providing specialized services within a region. These external agreements are normally formal, written plans that define the roles and responsibilities of the participants and can be categorized as automatic aid, mutual aid, or outside aid agreements.

Figure 1.23 In areas protected by volunteer and combination fire departments, response times may be increased as members arrive in their own vehicles.

Automatic Aid

Automatic aid is a formal, written agreement between jurisdictions that share a common boundary. The authority to request automatic aid is understood and the dispatch or communication center alerts the correct unit or jurisdiction. Automatic aid occurs whenever certain predetermined conditions occur.

Fire officers who respond to automatic aid incidents operate in the same manner as when they respond to any emergency incident. If the incident involves activities that place them under the operational control of another jurisdiction, the officer should perform the following:

- Determine the proper operational radio frequency
- Determine the location of the Command Post
- Report to the IC
- Adhere to the personnel accountability system that is used by the primary agency
- Adhere to the established procedures
- Maintain situational awareness

When an officer is in charge of an incident that will involve automatic aid units from other jurisdictions, the officer is responsible for the following:

- Requesting the dispatcher to instruct all responding units acknowledge that they are on the assigned radio frequency
- Assigning units based on arrival time and capabilities
- Establishing and communicating the location of staging areas

All fire officers should be familiar with the resources that will automatically respond to various target hazards or facilities and the circumstances under which automatic aid will be activated. It is a good idea to include these resources in the preplan for a particular building or target hazard.

Mutual Aid

Mutual aid is a reciprocal agreement between two or more fire and emergency services organizations. The agreements may be between local, regional, statewide, or interstate agencies that may or may not have contiguous boundaries. The agreement identifies the resources that will be provided, the types of incidents they will be used at, and how the actions of the resources will be monitored and controlled. Responses under a mutual aid agreement are usually an on-request basis. It is important to remember that these arrangements do not guarantee a response from outside organizations. For example, if the jurisdiction that receives the request has also been affected by a disaster, such as a tornado or hurricane, the request can be denied.

Outside Aid

Outside aid is similar to mutual aid except that payment rather than reciprocal aid is made by one agency to the other. Outside aid is normally addressed through a signed contract under which one agency agrees to provide aid to another in return for an established payment, which is normally an annual fee but which may be on a per-response basis. Otherwise, the outside agreement differs little from the mutual aid agreement.

Yourself

You must know your personal strengths and your weaknesses. Awareness of these characteristics allows you to apply your strong traits effectively during emergencies and decrease your weaknesses through training.

Strengths

A personal strength is the capacity or potential for effective action. Your strengths may be anything from a personality trait such as being tactful (diplomatic) to a physical skill such as setting a ground ladder. For a fire officer, strengths must include the ability to perform an accurate size-up and make decisions quickly based on current information for hazard mitigation. The ability to implement these decisions may be based on a personal strength in leadership or communication skills. Personal strengths may be divided into three categories: attitude, knowledge, and skill.

- *Attitude* — In today's fire service, actual fireground experience has decreased. In this constraint, personal attitude toward safety, learning, training, and constantly improving is important. If you have the right attitude, you can be an asset on the fireground. Your attitude will inspire people to follow you and take direction from you. You must have or develop the right attitude. For instance, consistently wearing a seat belt and enforcing seat belt policy combine to create an image of a safety conscious officer **(Figure 1.24)**.

Figure 1.24 Company officers must set the correct example for their subordinates by adhering to all department policies and procedures. For example by wearing seat belts.

- ***Knowledge*** — You use the information you have collected and stored in your brain to make decisions every second of your life. It can be divided into two types of knowledge, *factual* and *experiential*. Factual knowledge is learned through study, reading, and course work such as the content information in a hydraulics course. Experiential knowledge is gained through direct experience. Experience tells you what will work under certain conditions and what will not work. For instance, you may learn that the most effective means of extinguishing a compartment fire is to direct a straight stream of water onto the ceiling over the fire. If this works, you will probably use it again in a similar circumstance.

  Examples of knowledge that you must acquire and maintain include:
  — Building construction
  — Fire behavior
  — Strategy and tactics
  — Water flow and distribution
  — Hazard recognition
  — Local standard operating procedures/guidelines (SOPs/SOGs)
  — Locally adopted Incident Command System (ICS)

- ***Skills*** — Skills are physical activities. A skill consists of a series of steps that are taken, usually in a specific order, to complete a task or activity. A skill may be as simple as tying a knot to completing a complex series of steps to shore a trench. Skills are learned and practiced in training evolutions and computer-generated scenarios. Because current trends in emergency responses have resulted in a decrease in fire-related situations that can provide direct experience, it is important that training evolutions and computer simulations be as realistic as possible.

  Skills that are essential to being a successful IC include:
  — Communication skills
  — Decision-making
  — Ability to work under pressure
  — Size-up
  — Personnel supervision
  — Leadership, directing multiple unit operations
  — Delegation

Developing your knowledge and skills will help you:
— Predict fire behavior
— Predict fire development and spread in various occupancy types
— Predict the weakening effects fire development and spread will have on various types of structures
— Determine the volume of water required to control a confined (interior) structure fire
— Determine the correct strategy and tactics required to control and extinguish a fire

— Determine the strategic objectives and establish an Incident Action Plan
— Assign resources effectively
— Properly size-up a variety of types of situations

Weaknesses

It is important to know your weaknesses as well as your strengths. Acknowledging those weaknesses can motivate you to correct them through additional training, study, or behavior modification. However, it must be stressed that you should never attempt any action or activity that is beyond your ability. Nor should you commit limited resources to situations that are beyond you or your personnel's level of training or ability. This is especially true in situations involving offensive interior fire attack where sufficient water supply is not available or sufficient personnel to adhere to the **two-in/two-out rule** are not available.

One type of strength can actually become a weakness: overconfidence. Being self-confident can assure you that your decisions are correct based on the information available to you. However, overconfidence can cause you to take risks or place personnel in unsafe situations. Overconfidence can lead to faulty decisions that are based on personal perceptions and not on the actual conditions. The end result of overconfidence can be disastrous. Continuous training in simulated situations is an excellent tool for measuring your capabilities and deficiencies. Training allows you to build confidence in controlled situations that are close to reality.

> **Two-In/Two-Out Rule** — Occupational Safety and Health Administration (OSHA) regulation requiring a team of at least two personnel to be organized before entering an Immediately Dangerous to Life or Health (IDLH) atmosphere. It requires a standby team of at least two personnel outside the IDLH atmosphere to back up the entry team in the event they require rapid rescue.

Preincident Planning

One of the primary tools available to you as an IC is the preincident plan. Preincident planning allows you to anticipate the resources and tactics needed to control hazardous situations in specific structures or types of structures within your jurisdiction.

Because every emergency incident is an uncontrolled situation, they are rarely identical. While similarities to other emergencies may exist, the exact circumstances surrounding an emergency will likely be different. To ensure that you are able to recognize similarities and adapt to differences, it is important for you to know as much as possible about the potential incident scene before an emergency occurs.

Although gathering information about a particular structure or occupancy during a preincident survey is often referred to as preincident planning, gathering information is only one part of the process. Preincident planning is the entire process of:

- Gathering and evaluating information
- Creating or gathering drawings or aerial photos that show the lay of the land and water supplies
- Developing procedures based on that information
- Distributing the information to personnel and units that will respond to the site
- Ensuring that the information remains current

Target Hazard — Facility or site in which there is a high potential for life or property loss.

To obtain this information, officers and personnel conduct preincident surveys of commercial, industrial, multifamily residential, and institutional occupancies and high-risk **target hazards** within their response areas.

Preincident planning consists of the following four separate functions:

1. Developing positive relationships with building owners/occupants
2. Conducting the preincident survey
3. Managing preincident data
4. Developing preincident plans

Basic to the preincident planning process is an understanding of the difference between a preincident survey and a fire and life safety code enforcement inspection. While some jurisdictions authorize company officers to perform both types of activities, most do not. Preincident surveys are similar to fire and life-safety inspections, although surveys are not intended to locate code violations. If violations are discovered during a survey, the company officer may request that the owner/occupant correct the violation or the officer may simply report the problem to the authority's fire inspection and code enforcement division.

Company-level personnel conduct preincident surveys for the following reasons:

- Become familiar with a structure or facility, its physical layout and design, any built-in fire protection systems, water supply capabilities, and any hazards that may exist **(Figure 1.25)**.

- Visualize and discuss how an emergency is likely to occur or a fire may behave in an occupancy and how existing strategies and tactics might apply to an incident at this occupancy.

- Identify the number and types of resources needed to handle an incident at a particular location.

- Identify critical conditions that were not noted during any previous facility surveys or have changed since the last survey.

- Establish a proactive relationship with the building's owner/occupant.

Company officers and unit personnel must also have an understanding of basic building construction and the building codes that regulate construction in the jurisdiction. Building codes define the type of construction materials and techniques that are used to build structures that will be used for specific purposes. These specific purposes determine the type of occupancy classification that will be assigned to the completed structure. Company officers develop preincident plans based on the completed structure and its use and occupancy type.

You should also take advantage of the opportunity to survey buildings while they are under construction or renovation. These surveys provide opportunities to view and discuss various construction techniques and building components that will be hidden once those structures are complete **(Figure 1.26)**. At the same time, you can develop operational plans in the event that a fire occurs during the construction of the building. Photographs of the building under construction should be taken and added to the building survey file so future firefighters can see how the building was constructed.

You may find it helpful to refer to NFPA® 1620, *Standard for Pre-Incident Planning*. This document provides the information needed to conduct thorough preincident surveys and sample forms needed to develop comprehensive preincident plans from the information gathered during surveys. Some fire service organizations equip their apparatus with onboard computer systems that provide preincident information—this is an excellent tool to use while responding to an incident.

To be an effective tool, the preincident plan must have certain characteristics. The plan must be accessible, current, and accurate:

- ***Accessible*** — The plan must be easily and rapidly accessible to the IC during response to the incident. That is, the information must contain the most important information in an easy-to-read format. The information may be contained on one or two sheets of paper in a notebook, printed out with the dispatch information, or accessed through a wireless computer network **(Figure 1.27)**.

Figure 1.25 Preincident surveys permit firefighters to collect information about potential hazards within a structure.

Figure 1.27 Preincident plans may be accessed through computer terminals in the apparatus.

Figure 1.26 Surveys of new construction sites allow firefighters to see areas that will become hidden voids when the structure is completed. *Courtesy of Ron Moore.*

Chapter 1 • Managing the Incident **37**

- ***Current*** — Information must be up-to-date. All recent changes in the site must be included. This will require constant monitoring of all construction and renovation activities and annual or biannual surveys of the structures in the response area.

- ***Accurate*** — Data collected for the plan must be accurate and contain as much detail as possible. When the data is analyzed, the appropriate information can be included in the final plan while a site file can be created for future reference. The site file can be maintained separately from the preincident plan.

Information to be included in the plan should be a site plan, special hazards, and important data:

- ***Site plan*** — A site plan provides a visual reference to the location of the structure and its relationship to other structures, streets, and barriers. The site plan is created during the survey and is part of the final plan. The site plan should be easy to read and absorb quickly **(Figure 1.28)**. The response time may be the only time that you have to get a feel for the layout of the building. You may only have time to look at one diagram or picture before arrival. Too much detail can confuse and even distract you.

- ***Important data*** — The analysis of the data collected during the survey should result in a list of the most important details to be included on the plan.

- ***Special hazards*** — Commercial and industrial occupancies may contain special hazards that should be indicated on the plan. These hazards may include hazardous processes or materials, life safety hazards, access restrictions, or lack of water pressure or availability. The location of the facility's Safety Data Sheets (SDS) should be noted on the site plan.

When developing any preincident plan or performing a preincident survey, you must always conform to your organization's standard operating procedures. In most cases that will require the fire officer to be responsible for ensuring that all preincident plans are accurate and complete for facilities in their respective response area.

Summary

Effectively controlling any emergency incident is the result of good incident management. Incident management includes applying a locally adopted and adhered to Incident Command System, properly using a decision-making model, and having readily accessible information based on a preincident plan. As an Incident Commander, you must be able to make the right decisions under pressure based on knowledge of the incident site, your resources, and yourself. To accomplish this, you must know the difference between incident priorities, strategies, tactics, and tasks as well as how they apply to the general steps for making a decision-making model. The information in this chapter will provide a foundation for the remaining chapters and for the training scenarios provided in this manual.

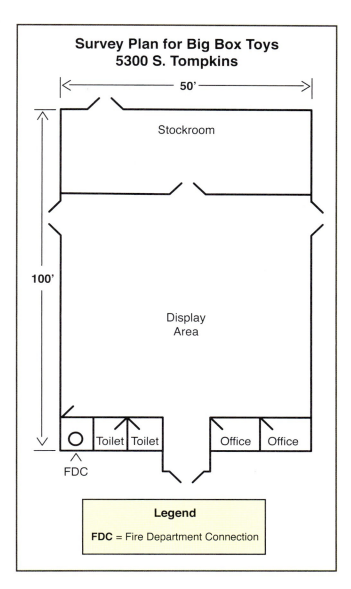

Figure 1.28 Example of the contents of a site plan.

Review Questions

1. What are the common characteristics of the Incident Command System (ICS)?
2. What are the basic organizational levels of the ICS?
3. What are the elements of personnel accountability?
4. What are the incident priorities?
5. What are the incident strategies?
6. What is the difference between tactics and tasks?
7. What are the steps of the decision-making process?
8. What things do you need to know about your local response area?
9. What are the most basic resources at your disposal at the incident?
10. What are the four functions of preincident planning?

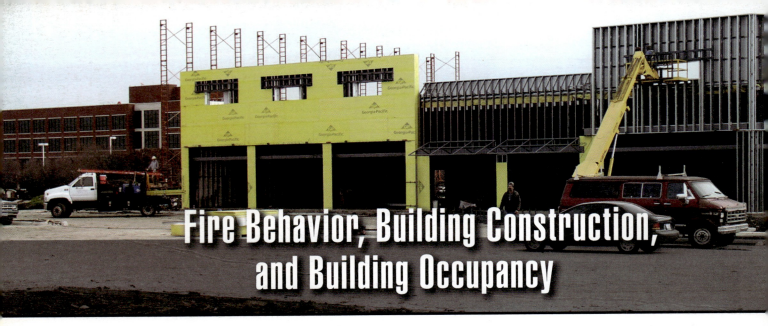

Fire Behavior, Building Construction, and Building Occupancy

Chapter Contents

Fire Behavior **44**
    Fire Spread 45
    Fire Behavior in Compartments 47
    Rapid Fire Development 48
    Factors That Affect Fire Development 54

Building Construction **60**
    Construction Type 63
    United States Construction 63
    Canadian Construction 69
    Building Periods or Age 70
    Interior Building Arrangement 72

    Basements, Cellar, and Crawl Spaces 73
    Attics and Cocklofts 75
    Concealed Spaces 76
    Structural Collapse Potential 76

Occupancy Types **83**
    Single Use 83
    Multiple Use 83

Building Contents **88**
Summary **89**
Review Questions **89**

Photo courtesy of Ron Moore.

chapter 2

Key Terms

Ambient Conditions	59
Autoignition Temperature (AIT)	48
Backdraft	51
Balloon Frame Construction	60
Bearing Walls	67
Collapse Zone	78
Compartment	45
Conduction	45
Convection	45
Fireproof	64
Fire Stop	65
Flashover	49
Fuel	45
Fuel Load	43
Heat	45
Occupancy Classification	43
Piloted Ignition	48
Platform Frame Construction	60
Plenum	76
Protected Steel	64
Pyrolysis	45
Radiation	45
Structural Collapse	76

Fire Behavior, Building Construction, and Building Occupancy

Learning Objectives

After reading this chapter, students will be able to:

1. Describe basic fire behavior.
2. Explain building construction as it relates to strategy and tactics.
3. Identify each of the building construction classifications.
4. Discuss occupancy types.
5. Discuss building contents.

Chapter 2
Fire Behavior, Building Construction, and Building Occupancy

Courtesy of Ron Moore.

Sun Tzu wrote in his book *The Art of War* that you must know your enemy. For firefighters that enemy is fire. Fire has its own characteristics and behavior. It can be harnessed and used for good or it can be uncontrolled and cause death and destruction. By knowing how fire behaves, you can predict how it will develop and spread allowing you to determine the most effective way to control, confine, and eliminate it as an uncontrolled hazard.

Because understanding basic fire behavior is important for all fire service personnel, it is a basic component of the Firefighter Level I training provided to entry-level personnel. As the initial Incident Commander, knowledge of fire behavior is extremely important in order to determine the correct strategy and tactics to apply to a compartment or structure fire. You must, therefore, know how fire behavior influences and is influenced by the environment in whch it can occur.

For instance, if your jurisdiction contains only grasslands, then you must know the types of vegetation in the area, how they will support the ignition and growth of fire, the weather patterns that will aid or hinder fire control, and the topography that the fire will spread over. If your jurisdiction is primarily urban, containing a variety of types of structures, you must understand building construction, including:

- How buildings are classified by the building codes
- The types of building materials used
- The general interior arrangement of buildings
- How fire will behave within a structure

Building construction is only part of the fire environment that you must consider. You must also know the uses of the buildings in your response area. Building use is regulated by locally adopted building, fire, and life safety codes that specify the types of processes and activities that can occur in specific types of buildings. Referred to as **occupancy classifications**, these codes place similar types of activities into similar classes **(Figure 2.1, p. 44)**. In an average municipality or jurisdiction, there will be at least one of each type of occupancy classification present.

Finally, it is important that you be able to recognize how building contents, referred to as the **fuel load**, can contribute to a structure fire. Most fires in structures initially involve the contents before spreading to the structure

Occupancy Classification — Classifications given to structures by the model building code used in that jurisdiction based on the intended use for the structure.

Fuel Load — The total quantity of combustible contents of a building, space, or fire area, including interior finish and trim, expressed in heat units or the equivalent weight in wood.

Figure 2.1 Modern strip shopping malls are vastly different in appearance from those built in the previous fifty years. *Courtesy of Ron Moore.*

itself. In the early stage of a fire, an offensive attack can quickly extinguish a contents fire and prevent it from spreading. However, contents fires that go undetected for many hours will consume all the oxygen in a structure creating the potential for an explosive condition when fresh air is introduced.

In this chapter, you will learn about fire behavior, building construction, occupancy types, and building contents.

Fire Behavior

In this manual, the discussion of fire behavior is restricted to those fires that start in a compartment or confined space of a structure. The compartment may range in size from a small utility closet to one that consists of the entire interior of the structure without interior walls or partitions, such as a warehouse or retail store. In any situation, how the fire develops and spreads will be determined by the available fuel, oxygen, and the structural configuration of building components such as the floors, walls, and roof or ceiling assemblies.

Theories of fire growth are generally based on models developed and tested in controlled environments in testing laboratories. In those laboratories the various stages of fire growth, the time and temperature relationships, and the definitions of extraordinary fire events have been tested and proven. Like most laboratory experiments, as long as all the conditions are constant, the results will generally be the same **(Figure 2.2)**.

Figure 2.2 In an effort to determine how fires behave, the National Institute of Standards and Technology performs a variety of test fires under controlled conditions. *Courtesy of National Institute of Standards and Technology.*

As a firefighter and a fire officer, you know from experience that fire behavior does not always conform to the results found in laboratories. Therefore, you must take the theories of fire behavior and use them as a baseline to evaluate and predict actual fire behavior and to compare them with your experiences and perceptions.

In this portion of the chapter, fire behavior is discussed including fire spread and rapid fire development such as flashover, backdraft, and smoke explosion or fire gas ignition. Finally, the differences between structure fires and content fires are discussed. This section provides the foundation for the following sections on building construction, occupancy types, and contents.

Fire Spread

As long as there is sufficient **fuel** and oxygen, a fire will continue to grow and spread. Fire spread results from the concept of heat transfer which is basic to fire behavior. **Heat** will transfer from a burning object to other objects of lower temperatures thus spreading the fire.

Heat moves from warmer objects to cooler objects. The rate at which heat is transferred is related to the temperature differential of the objects and the thermal conductivity of the material of which the objects are made. For any given material, the greater the temperature differences between the objects, the more rapid the transfer rate. As the temperatures of the two objects get closer to each other, the rate of transfer will slow down. Heat can be transferred from one object to another by three methods: **conduction**, **convection**, and **radiation** (Figures 2.3a-c, p. 46).

Fire spread is controlled by whether the fire is unconfined or confined. When a fire occurs in an unconfined area, such as an outdoor campfire, much of the heat produced by the fire dissipates into the atmosphere through radiation and convection. However, when the fire is confined within a **compartment**, the walls, ceiling, floor, and objects in the compartment absorb the radiant heat produced by the fire. Once heat levels rise to a degree where those objects are at the same temperature, radiant heat energy is reflected back, continuing to increase the temperature of other fuel in the space increasing the rate of **pyrolysis** of the fuel and the rate of combustion. Hot air and smoke heated by the fire become hotter than the surrounding air, thus making them more buoyant and causing them to rise. Upon contact with cooler materials, such as the ceiling and walls of the compartment, heat is transferred to the cooler materials, raising their temperature. This heat transfer process raises the temperature of all materials in the compartment. As nearby fuel is heated, it pyrolizes. Eventually the rate of pyrolysis can reach a point where flaming combustion can be supported and the fire spreads.

When sufficient oxygen is available, fire development is controlled by the characteristics and configuration of the fuel. Under these conditions, the fire is said to be fuel controlled. This is why knowing the contents or fuel load is important in your decision-making process and in predicting fire growth. As a fire develops within a compartment, the fire reaches a point where further development is limited by the available oxygen supply and the fire is said to be ventilation controlled (**Figure 2.4, p. 46**). Fire officers and firefighters must realize that any action they take to gain access, create openings, or penetrate walls will cause a rapid increase in the amount of oxygen in the compartment. This can lead to rapid fire development.

Fuel — Flammable and combustible substances available for a fire to consume. Fuel can be in a solid, liquid, or gas form. In terms of a structure fire, fuel includes the building and its contents.

Heat — Form of energy associated with the motion of atoms or molecules in solids or liquids that is transferred from one body to another as a result of a temperature difference between the bodies such as from the sun to the earth. To signify its intensity, it is measured in degrees of temperature.

Conduction — Physical flow or transfer of heat energy from one body to another through direct contact or an intervening medium from the point where the heat is produced to another location or from a region of high temperature to a region of low temperature.

Convection — Transfer of heat by the movement of heated fluids or gases; usually in an upward direction.

Radiation — The transmission or transfer of heat energy from one body to another body at a lower temperature through intervening space by electromagnetic waves such as infrared thermal waves, radio waves, or X rays. Also called Radiated Heat.

Compartment — A room or space within a building or structure which is enclosed on all sides, at the top and bottom. The term *compartment fire* is defined as a fire that occurs within such a space.

Pyrolysis — Thermal or chemical decomposition of fuel because of heat that generally results in a lowered ignition temperature of the material. The preignition combustion phase of burning during which heat energy is absorbed by the fuel, which in turn gives off flammable tars, pitches, and vapors.

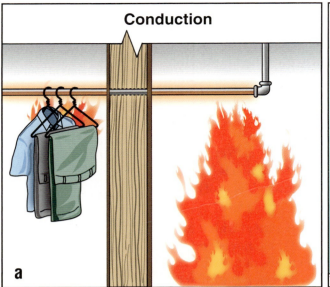

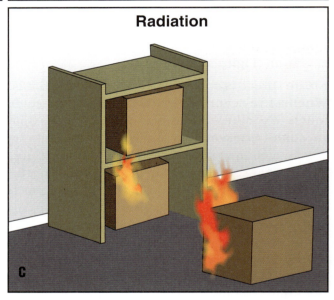

Figures 2.3a-c Compartment fires may spread through (a) conduction, (b) convection, or (c) radiation or a combination of these transfer methods.

Figure 2.4 Comparison of fuel and ventilation controlled fires.

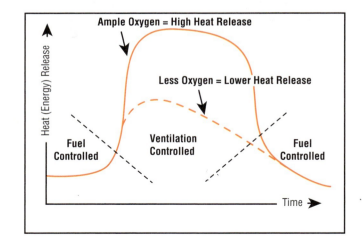

Fire Behavior in Compartments

When a fire develops in a compartment, heated products of combustion and entrained air become hotter than the surrounding air and rise to the ceiling in a plume. When these hot gases reach the ceiling they mushroom, spreading horizontally through the compartment **(Figure 2.5)**. The gases continue to spread until they reach the walls of the compartment. As combustion continues, the depth of the gas layer then begins to increase. The difference between hot smoke and the cooler air below causes them to separate into two distinct thermal layers.

The thermal layering of gases is the tendency of gases to form into layers according to temperature. The hottest gases will be in the top layer, while the cooler gases form the lower layer. In addition to the effects of heat transfer through radiation and convection, radiation from the hot gas layer also acts to heat the interior surfaces of the compartment and its contents. Thermal layering is sometimes referred to as heat stratification.

As the volume and temperature of the hot gas layer increases, so does the pressure. Like water, gas will expand when heated. Higher pressure in this hot layer causes it to push down and out. As the layer of gas at the ceiling level increases in temperature, it fills and increases the pressure in the compartment. The thermal layer fills the room from the top down within the compartment and out through any openings such as doors or windows. Once in an opening it will try to rise, the temperature and rate of heat transfer will dictate how fast the layer will move and spread out from the compartment of origin.

Within a compartment, the pressure of the smoke and gas can affect the movement of air in or out of the compartment. Because the pressure of the cool gas layer is lower, air from outside the compartment is drawn into the compartment. In the hotter gas layer near the ceiling, pressure is higher, thus forcing smoke out openings. At the point where these two layers meet as the hot gases

Figure 2.5 A smoke column will spread horizontally when it reaches the ceiling of the room.

exit through an opening, the pressure is neutral. The interface of the hot and cooler gas layers at the opening is commonly referred to as the neutral plane **(Figure 2.6)**. This neutral pressure may be visible at openings where hot gases are exiting and cooler air is moving into the compartment. It may also be visible in hallways between the point of ventilation and room fire.

During the development of a compartment fire, pyrolysis of exposed fuels can produce combustible gases, which can gather at locations in the layer remote from the fire plume. These pockets of gas may undergo **piloted ignition** from the transfer of heat energy directly from the fire plume itself or ignition as a result of having reached their **autoignition temperature**.

Rapid Fire Development

Over the years, the fire service has coined words and phrases to describe various fire conditions that have resulted from rapid fire development. Among these conditions are:

- Flashover
- Backdraft
- Rollover
- Smoke explosion
- Autoignition

> **Piloted Ignition** — Occurs when a mixture of fuel and oxygen encounter an external heat source with sufficient heat energy to start the combustion reaction.

> **Autoignition Temperature (AIT)** — The temperature to which the surface of a substance must be heated for ignition and self-sustained combustion to occur. The autoignition temperature of a substance is always higher than its piloted ignition temperature.

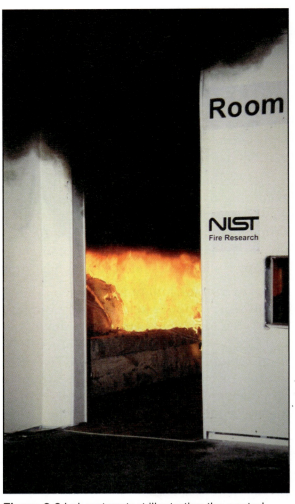

Figure 2.6 Laboratory test illustrating the neutral plane of a fire. *Courtesy of National Institute of Standards and Technology.*

Rapid fire development, sometimes referred to as extreme fire behavior, has been responsible for numerous firefighter deaths and injuries. Between 1990 and 2000, over 50 firefighters were killed as a result of rapid fire development. To protect yourself and your crew, you must be able to recognize the indicators of rapid fire development, know the conditions created by each of these situations, and determine the best action to take when they occur. In this section, each of these conditions will be discussed along with the general indicators that you should look for.

Indicators

Each of the types of rapid fire behavior have similar indicators, which identify that rapid fire development is imminent. The indicators are:

- Smoke rapidly exiting doors, windows, or other openings
- Doors forced open into the structure as fresh air is rapidly drawn in
- Smoke under pressure pulsing out of openings
- Heavily smoke-stained or cracked window glass
- Rapid lowering of the neutral plane
- Rapid rising and lowering of the smoke layer
- Rapid change in smoke color to black
- Rapid change in temperature within the compartment
- Yellow or orange flames at the ceiling moving away from the main body of fire
- Smoke being sucked back into the structure

The appearance of any of these indicators should cause you to reevaluate your tactics. Notice the various indicators listed in each of the following types of rapid fire development.

Flashover

Flashover occurs when all surfaces and objects within a compartment have been heated to their ignition temperature and ignite almost simultaneously. Flashover conditions are defined in a variety of ways; however, during flashover, conditions in the compartment change very rapidly from partial to full involvement of the compartment. When flashover occurs, burning gases push out of openings in the compartment (such as a door leading to another room) at a substantial velocity **(Figure 2.7)**.

> **Flashover** — A transitional phase in the development of a compartment fire in which surfaces exposed to thermal radiation reach ignition temperature more or less simultaneously and fire spreads rapidly throughout the space resulting in full room involvement or total involvement of the compartment or enclosed area.

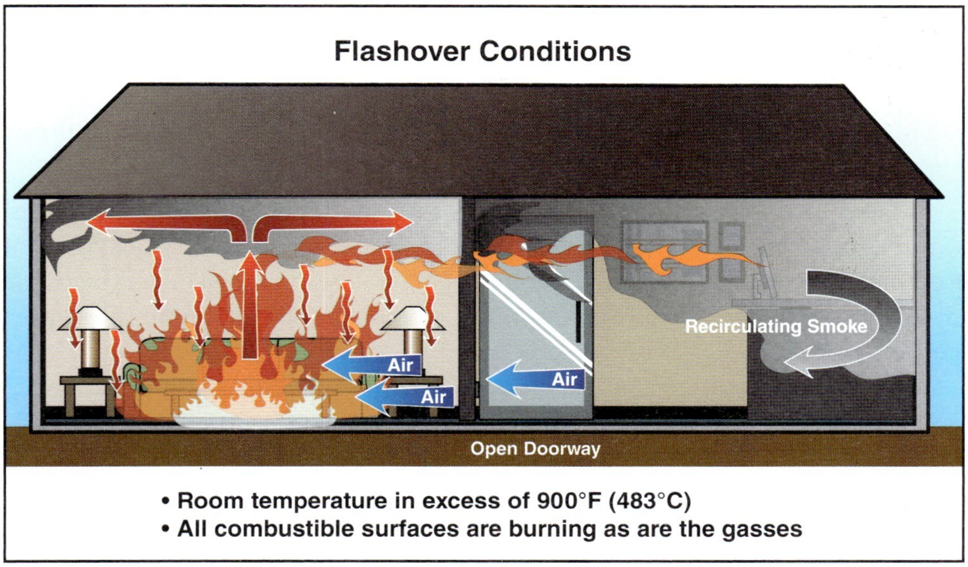

Flashover Conditions

- Room temperature in excess of 900°F (483°C)
- All combustible surfaces are burning as are the gasses

Figure 2.7 An example of conditions that may exist during flashover.

Just before flashover occurs, several things happen within the burning compartment:

- Temperatures rapidly increase from floor to ceiling.
- Additional fuel becomes involved.
- There may be extremely low visibility in the compartment area.
- The fuel in the compartment is producing combustible gases through pyrolysis.

As flashover occurs, the combustible materials in the compartment and the gases produced by pyrolysis ignite almost simultaneously; the result is full-room involvement. Survival rates for rescuers and occupants are extremely low in a flashover condition.

Elements of Flashover

There are four common elements of flashover:

- **Transition in fire development** — Flashover represents a transition from the growth stage to the fully developed stage.
- **Rapidity** — Though not an instantaneous event, flashover happens rapidly, in a matter of seconds, to spread complete fire involvement within the compartment.
- **Compartment** — There must be an enclosed space or compartment such as a single room or enclosure. There must be an upper level restriction of the developing gas layer, such as a ceiling.
- **Ignition of all exposed surfaces** — Virtually all combustible surfaces existing in the enclosed space become ignited.

Flashover does not occur in every compartment fire. Two factors determine whether a fire within a compartment will progress to flashover. First, the fuel must generate enough heat energy to develop flashover conditions. Characteristics of the fuel, including the surface area, its combustibility, and size of the compartment will contribute to the flashover potential. The second factor is ventilation. A developing fire must have sufficient oxygen to reach flashover. If there is insufficient ventilation, the fire may enter the growth stage but not reach the peak heat release of a fully developed fire.

When ventilation is increased (i.e., due to failure of window glazing or firefighters opening a door or window), additional oxygen will increase the fire and heat release. Remember, the amount of heat being released by the fire is directly proportionate to the amount of oxygen that is available to support combustion.

It is important to recognize that most fires that grow beyond the incipient stage become ventilation controlled. When windows are intact and doors are closed, the fire may move into a ventilation-controlled state more quickly. While this reduces the heat release rate, fuel will continue to pyrolize, creating extremely fuel-rich smoke. Today's building codes create tight, energy-efficient buildings that do not vent as rapidly as older buildings. This heat retention results in compartment temperatures rising quickly and a rapid progression to flashover, which in turn makes careful ventilation critical.

Backdraft

A ventilation-controlled compartment fire can produce a large volume of flammable smoke and other gases due to incomplete combustion. This mixture of flammable products can be heated well above its upper flammable limit. While the rate of heat release from a ventilation-controlled fire is limited, elevated temperatures are usually still present within the compartment. Although rare, an increase in low level ventilation (such as opening a door or window) prior to upper level ventilation can result in an explosively rapid combustion of the flammable gases, called a **backdraft**. Backdraft occurs in the decay stage, in a pressurized, unvented space that lacks oxygen **(Figure 2.8)**.

When potential backdraft conditions exist in a compartment, the compartment is filled with unburned fuel gases and smoke that are at or above their ignition temperature and only lacking sufficient oxygen to burn. A backdraft

Backdraft — The explosive or rapid burning of heated gases that occurs when oxygen is introduced into a compartment that has a depleted supply of oxygen due to an existing fire

Backdraft Sequence

Conditions:
- Low oxygen
- High heat
- Smoldering fire
- High fuel vapor concentrations

- Fresh air enters compartment
- Fuel vapor and air mix

- Vapor/air mixture reaches explosive limit
- Backdraft occurs

Figure 2.8 An example of conditions that may exist during backdraft.

can occur without the creation of a horizontal opening. All that is required is the mixing of hot, fuel-rich smoke with air. Backdraft conditions can develop within a room, a void space, or an entire building. Anytime a compartment or space contains hot combustion products, potential for backdraft must be considered before creating any openings into the compartment. To some degree, the violence of a backdraft is dependent on the extent to which the fuel/air mixture is confined. The more confined, the more violent it will be.

As with flashover, it is critical to recognize the potential warning signs of backdraft conditions. Common indicators of the potential for a possible backdraft include:

- Intact smoke-stained or crazed windows indicating the existence of an under-ventilated fire
- Smoke pushing from under eaves indicating a heavy pressure condition within the structure
- Smoke pulsing rapidly in and out of exit points on the structure
- Air rushing into the structure or compartment when a horizontal opening is made
- A swirling pattern in the smoke at floor level as air is sucked into the structure or compartment through a door or window; no clear interface between the smoke and fresh air is visible
- Blue or dancing flames along the ceiling that are detached from the main body of fire
- Numerous tiny mushroom shaped puffs of smoke exiting the compartment or structure through a doorway or window
- A whistling or roaring sound as fresh air is drawn into the compartment or structure through a door or window

It is often incorrectly assumed that a backdraft will always occur immediately after making an opening into the building or involved compartment. Mixing of hot flammable products of combustion with air through the action of gravity, air current, pressure differential, and wind effects sometimes takes time so that backdraft may not occur until several moments after air is introduced. You must watch what the smoke is doing: air current changing direction, neutral plane lifting, or smoke rushing out are all indicators of extreme fire behavior.

The effects of a backdraft can vary considerably depending on a number of factors including:

- Volume of smoke
- Degree of confinement
- Pressure
- Speed with which fuel and air are mixed
- Location where ignition occurs

Rollover

The term *rollover* describes a condition where the unburned fire gases accumulated in the top layer of fire gases in a compartment ignite and flames spread across the ceiling. Rollover is a fire gas ignition and also a significant indicator of impending flashover. Rollover is distinguished from flashover by its involvement of only the fire gases at the upper levels of the compartment and not the other fuel items within a compartment. Rollover may occur during the growth stage as the hot-gas layer forms at the ceiling of the compartment **(Figure 2.9)**. Flames may be observed in the layer when the combustible gases reach their ignition temperature. While the flames add to the total heat generated in the compartment, this condition is not flashover. Rollover may precede flashover, but will not always result in flashover.

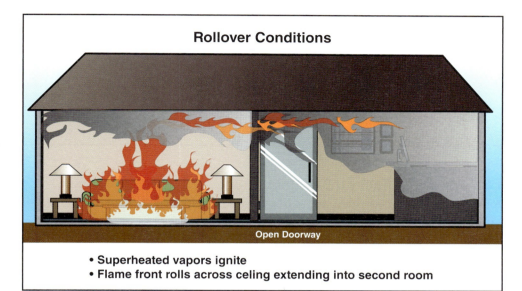

Figure 2.9 An example of conditions that may exist during rollover.

Smoke Explosion

While backdraft is considered a pressure induced force behavior, a smoke explosion is a non-pressurized rapid fire behavior. A smoke explosion may occur pre- or post-decay and is unburned fuel gases in search of an ignition source. When smoke travels from the fire it can accumulate in areas or exit from openings. The smoke contacts an ignition source and results in an explosively rapid combustion called a smoke explosion. Smoke explosions are usually not as violent as a backdraft because they do not occur in a pressurized compartment **(Figure 2.10)**.

Autoignition

Autoignition occurs when smoke and flammable gases are above their upper flammable range and have temperatures higher than their ignition temperatures. The smoke and gases then exit a compartment and mix with fresh air to ignite spontaneously. This condition may also resemble a smoke explosion if the ignition of smoke and gases burns back into the compartment. Autoignition may occur at various points including windows, doors, and stairways **(Figure 2.11, p. 54)**.

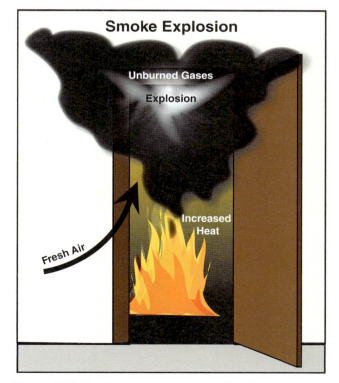

Figure 2.10 Smoke explosions occur when heated smoke mixes with fresh air in the presence of an ignition source.

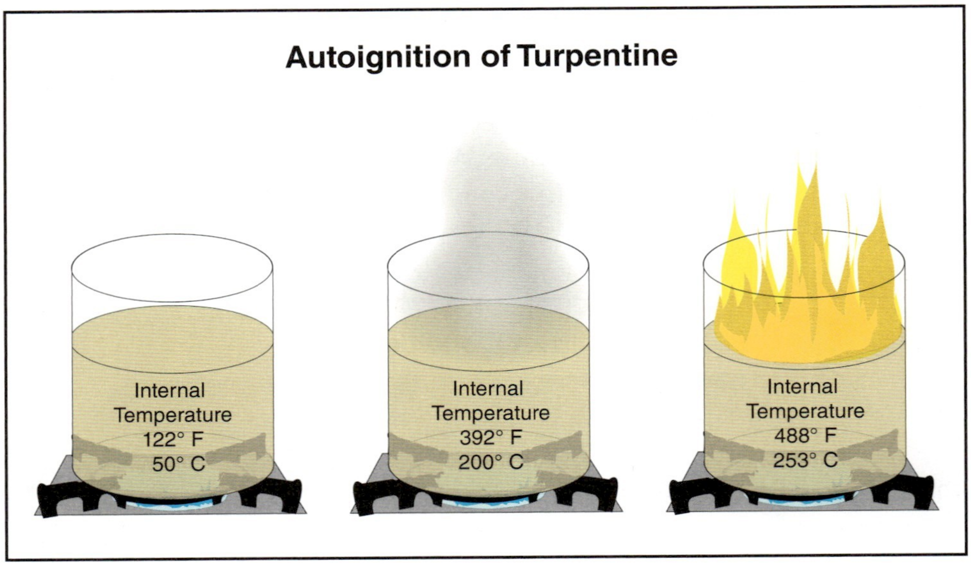

Figure 2.11 Autoignition occurs when a fuel is heated to the point that it ignites spontaneously.

Autoignition of Turpentine

Internal Temperature 122° F / 50° C

Internal Temperature 392° F / 200° C

Internal Temperature 488° F / 253° C

Factors That Affect Fire Development

A number of factors influence fire development within a compartment. These factors include the following:

- Fuel type and amount of surface exposure
- Availability and location of additional fuel in relation to the fire location
- Compartment volume and ceiling height
- Ventilation and changes in ventilation
- Thermal properties of the compartment
- Fuel load

Fuel Type

The type of fuel involved in combustion affects both the amount of heat released and the time over which the release occurs. Fires involving all of the following fuels will eventually spread to the contents and structure of the compartment resulting in a primarily Class A fueled fire.

Fuel types include:

- Class A or cellulose type fuels are the most common types of fuels found in structures.
- Class B type fuels, flammable/combustible liquids and gases, may exist in limited quantities in some occupancies (such as residences) and in large quantities in other occupancies (such as paint supply stores).
- Class C type fuels, energized electrical wiring, equipment, and appliances, may provide an ignition source but rapidly expand into a Class A or B fire as the fire spreads to adjacent fuel sources. Remember that if the fuel is not energized, it is not a Class C fire.
- Class D type fuels, combustible metals, are a hazard found in limited commercial or industrial occupancies and vehicles.
- Class K type fuels, deep fat fryers, are mainly associated with commercial kitchens.

In a compartment fire, the most fundamental Class A fuel characteristic influencing fire development is surface-to-mass ratio. Combustible materials with high surface-to-mass ratios are much more easily ignited and will burn more quickly than the same substance with less surface area. Therefore, sawdust in a lumber mill will ignite and burn more rapidly than the raw logs stored there.

Compartment fires involving Class B flammable/combustible liquids will be influenced by the surface area and type of fuel involved. A liquid fuel spill will increase that liquid's surface to volume ratio and will generate more flammable vapors than that same liquid in an open container **(Figure 2.12)**. The increase of vapor due to the spill will also allow for more of the fuel to ignite, resulting in greater heat over a shorter period of time.

It is important to remember that today's homes and businesses are largely filled with contents made from petroleum-based materials. Black smoke is often present in fires involving materials like foam, synthetics, and plastics.

A compartment fire that results from a flammable/combustible gas leak will begin with a rapid ignition of the gas and an explosion. If the fuel source is not controlled, it will continue to burn at the point of release extending to adjacent combustibles. Shutting off the fuel source or controlling the leak will reduce or eliminate the Class B fuel, but the resulting Class A fire will continue to burn until extinguished.

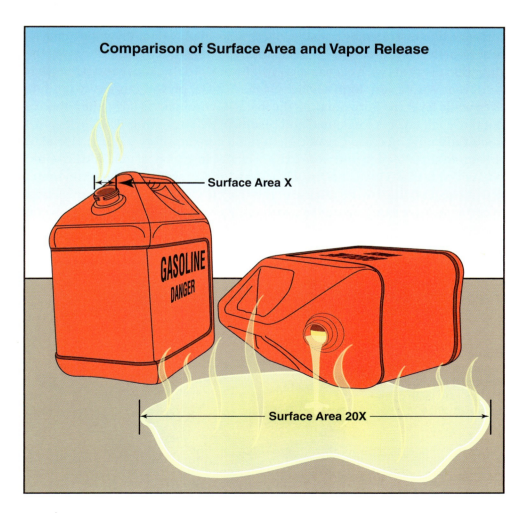

Figure 2.12 As a liquid's surface area increases, so does its vapor release rate.

Chapter 2 • Fire Behavior, Building Construction, and Building Occupancy

Availability and Location of Additional Fuel

Factors that influence the availability and location of additional fuels include the building configuration, construction materials, contents, and proximity of the initial fire to these exposed fuel sources.

Building configuration is the layout of the structure including:

- Number of stories above or below grade
- Compartmentation
- Floor plan
- Openings between floors
- Continuous voids or concealed spaces
- Barriers to fire spread

Each of these elements may contribute to fire spread or containment. For instance, an office with an open floor plan may contain furnishings that provide fuel sources on all sides of a point of ignition **(Figure 2.13)**. On the other hand, a compartmentalized configuration may provide fire-rated barriers, such as walls, ceilings, and doors, that separate fuel sources and limit fire development to an individual compartment.

All construction materials are affected by fire and many contribute to the fuel load of some types of buildings. For example, in wood frame buildings, the structure itself is a source of fuel. The orientation of these fuels as well as their surface-to-mass ratio will also influence the rate and intensity of fire spread. This can be illustrated by considering the use of plywood. Plywood consists of several thin layers of wood veneer laminated together while alternating the direction of grains with each layer until the desired thickness and strength is achieved. With a very high surface-to-mass ratio, plywood is easily ignited, even while level and horizontal. As the material continues to burn, the adhesive bond of each layer is weakened and results in delaminating. The surface-to-mass ratio now increases, resulting in rapid consumption of the material. If this same sheet of plywood were to be oriented in a vertical position, it would burn even faster. This is because the growing fire would preheat the material above it and cause the release of flammable vapors through pyrolysis. In addition to structural members, combustible interior finishes, such as wood paneling, can be a significant factor influencing fire spread.

Figure 2.13 Carpet, fabric panels, and decorative wood and plastic add to the fuel load in an open floor plan office.

Combustible Interior Finishes and Fire Spread

Combustible interior finishes have been a significant factor in a number of major fires that resulted in high fatalities and injuries. One of these fires was the Cocoanut Grove fire in Boston, Massachusetts, November 28, 1942. A total of 492 people lost their lives in this rapidly developing nightclub fire. Flammable decorations quickly spread the fire. Thick smoke and heat killed the occupants as they attempted to escape through the limited number of exits that were available.

Similarly, at 0705, November 21, 1980, a fire that began in the deli of the MGM Grand Hotel in Paradise, Nevada, spread smoke though the casino and up into the 26-story hotel. The result was the second worst hotel fire in U.S. history, killing 87 and injuring 650. The fire cause investigation indicated that cellulose panels applied with highly flammable adhesive as well as decorations made from synthetic and plastic materials produced high levels of heat and toxic gases.

Just over 51 years after the Cocoanut Grove fire, 100 people lost their lives in another New England nightclub fire. Pyrotechnics ignited combustible interior finish materials (polyurethane foam sound insulation) at the Station nightclub in West Warwick, Rhode Island, resulting in extremely rapid fire development which trapped many of the buildings occupants as they attempted to escape through the limited number of exits that were available.

The contents of a structure are often the most readily available fuel source, significantly influencing fire development in a compartment fire. When contents release a large amount of heat rapidly, both the intensity of the fire and speed of development will be increased. For example, synthetic furnishings, such as polyurethane foam, will begin to pyrolize rapidly under fire conditions due to the chemical makeup of the foam and its surface/mass ratio (even when located some distance from the origin of the fire), speeding the process of fire development.

The proximity, in relation to the fire, and continuity of contents and structural fuels also influences fire development. Fuels located in the upper level of adjacent compartments will pyrolize more quickly from the effect of the hot gas layer. Continuous fuels such as combustible interior finishes will rapidly spread the fire from compartment to compartment. Similarly, the location of the fire within the building will influence fire development. When the fire is located low in the building, such as in the basement or on the first floor, convected heat will cause vertical extension through atriums, unprotected stairways, vertical shafts, and concealed spaces. Fires originating on upper levels generally extend downward much more slowly through structural collapse or explosions.

Compartment Volume and Ceiling Height

All other factors being equal, a fire in a large compartment will develop more slowly than one in a small compartment. Slower fire development is due to the greater volume of air and the increased distance radiated heat must travel from the fire and contents that must be heated. Remember, though, that this large volume of air will support the development of a larger fire before the lack of ventilation becomes the limiting factor **(Figure 2.14, p. 58)**.

Figure 2.14 The current trend in mercantile structures includes high ceilings, exposed roof support structures, and large quantities of merchandise stored on shelves in the open.

A high ceiling may also mask the extent of fire development by allowing a large volume of hot smoke and other fire gases to accumulate at ceiling level, while conditions at floor level remain relatively unchanged. This situation is particularly hazardous because conditions can change rapidly if this hot gas layer ignites.

Ventilation

Ventilation in a compartment significantly influences how fire develops and spreads within the space. Pre-existing ventilation is the actual and potential ventilation of a structure based on structural openings, construction type, and building ventilation systems. For the most part, all buildings exchange air inside the structure with the air outside the structure. In some cases this is due to constructed openings such as windows, doors, and passive ventilation devices as well as leakage through cracks and other gaps in construction. In other cases, this air exchange is primarily through the heating, ventilating, and air conditioning (HVAC) system.

When considering fire development, it is important to consider potential openings that could change the ventilation profile under fire conditions. Under fire conditions, windows can fail or doors can be left open, increasing ventilation into the compartment. When a fire develops to the point where it becomes ventilation controlled, the available air supply will determine the speed and extent of fire development and the direction of fire travel. Fire will always grow in the direction of ventilation openings as the fire seeks fresh air **(Figure 2.15)**.

Figure 2.15 Fire always seeks a source of fresh air.

Thermal Properties of the Compartment

Thermal properties of a compartment include:

- **Insulation** — contains heat within the compartment causing a localized increase in the temperature and fire growth
- **Heat reflectivity** — increases fire spread through the transfer of radiant heat from wall surfaces to adjacent fuel sources
- **Retention** — maintains temperature by slowly absorbing and releasing large amounts of heat

The thermal properties of the compartment can contribute to rapid fire development resulting in flashover, backdraft, or smoke explosions. The thermal properties can also make extinguishment more difficult and reignition possible.

Ambient Conditions

While ambient temperature and humidity outside the structure can have an effect on the ignitability of many types of fuels, these factors are less significant inside a compartment. **Ambient conditions** such as high humidity and cold temperatures can slow the natural movement of smoke. Strong winds can significantly influence fire behavior by placing pressure on one side of a structure and forcing smoke and fire out the opposite side. If a window fails or a door is opened on the windward side of a structure, fire intensity and spread can increase significantly creating a blowtorch-like effect. During fire suppression activities, wind direction and velocity can prevent or assist in ventilation activities. Cold temperatures can influence size-up by causing smoke to appear white and giving the wrong impression of the interior conditions. Atmospheric air pressure can also cause smoke to remain close to the ground obscuring visibility during size-up **(Figure 2.16)**.

> **Ambient Conditions** — Common, prevailing, and uncontrolled atmospheric weather conditions. The term may refer to the conditions inside or outside of the structure.

Figure 2.16 Weather conditions may cause smoke to be concentrated near ground level. *Courtesy of Dick Giles.*

Fuel Load

The total quantity of combustible contents of a building, space, or fire area is referred to as the *fuel* load (some documents may use the term *fire* load). The fuel load includes all furnishings, merchandise, interior finish, and structural components of the structure **(Figure 2.17)**. Using a set of mathematical equations, a fire safety engineer can generate a fairly accurate estimate of the fuel load of any structure. However, as a fire officer, you may only be able to generate an estimate based on your knowledge and experience. For instance, a concrete block structure containing stored steel pipe will have a much smaller fuel load than a wood frame structure used for storing flammable liquids. Your knowledge of building construction and occupancy types will be essential to determining fuel loads.

Figure 2.17 In a modern department store, merchandise, counters, display shelves, and interior finish contribute to the fuel load. *Courtesy of Ron Moore.*

Building Construction

Over the past 200 years, devastating fires in urban areas in North America have resulted in the creation of building, fire and life safety, and zoning codes intended to limit the ignition and spread of fires. The majority of local jurisdictions (municipalities, counties, states, and provinces) have adopted either national model building codes or locally developed building codes that are based on the national models. In either case, these building codes regulate and define the types of construction materials and techniques that may be used and how the completed structure may be used.

The original date a building was constructed will have a significant influence on whether it meets a particular building code requirement. In most cases, existing structures are not required to meet current building codes unless significant alterations are proposed for the structure. Most existing buildings are exempted from requirements of the new code unless there is a retroactive clause adopted. Therefore, older buildings may not meet the same level of safety as new construction buildings. The best example would be mandatory sprinklers in buildings over three stories. All newer buildings would have to be built to that standard. Unless mandatory retrofits are required, all buildings constructed before the code was adopted would not have sprinklers. There are many examples of fire and life safety measures in the building codes to protect occupants in case of fire such as fire-resistant doors, self-closing doors, number and placement of exits, exit design, and fire and smoke resistant stair enclosures, in addition to sprinklers and standpipes.

It is essential to know the general construction types and materials used in your response area. Due to climate and geographic differences, building construction varies across much of North America. Northern houses may have 2 x 6 inch (50 mm by 150 mm) wall studs with asphalt roof shingles whereas houses in the southwestern United States may be commonly made with stucco siding over 2 x 4 inch (50 mm by 100 mm) frames and clay tiled roofs. Additionally, the age of structures and the time period that communities were developed and settled will also have an influence on the type of materials used in building construction.

Many residential neighborhoods may have a consistent construction design based on the time they were developed. Buildings built before the turn of the nineteenth century were mostly mortise and tenon, also called post and beam or brace frame. With the advent of lumber mills that could cut 2 x 4 inch (50 mm by 100 mm) lumber and the introduction of cheap, mass produced nails, **balloon frame construction** became common between the late 1800s and the mid-1950s. The speed and ease of building made balloon frame the desired design type for residential construction.

After World War II, **platform frame construction** became popular. The lack of long wood beams and the speed at which platform buildings could be built replaced balloon frame structures. Most houses currently being built contain some lightweight construction including trusses, foam sheathing, steel studs, or wood I-beams. The use of steel studs has influenced an increase of balloon framing once again. Unlike previous balloon frame construction, modern codes require fire-stopping measures which inhibit the spread of fire vertically within walls **(Figures 2.18a&b)**.

Balloon Frame Construction — A construction method using long continuous studs that run from the sill plate (located on the foundation) to the roof eave line. All intermediate floor structures are attached to the studs. Requires the use of long lumber and generally lacks any type of fire stopping within the wall cavity.

Platform Frame Construction — A construction method in which a floor assembly creates an individual platform that rests on the foundation. Wall assemblies the height of one story are placed on this platform and a second platform rests on top of the wall unit. Each platform creates fire stops at each floor level which restricts the spread of fire within the wall cavity.

In residential and commercial construction, areas developed by the same contractor will generally possess the same construction materials, designs, and characteristics. Fire company surveys of new construction are essential to understanding how the completed structure will perform during a fire.

While there is a general consistency in construction types based on the time frame the area was developed, this is not always true. In some urban areas, older homes are being torn down and replaced by new construction **(Figure 2.19)**. New construction that meets current building code requirements, architectural design, and building styles, may include condominiums, town houses, and multistory "McMansions." In many cases, building lots that formerly contained one residence now contain two or more with reduced building separations.

You must also be familiar with the interior arrangement of structures in your area. As mentioned in the fire behavior section, the interior arrangement of a compartment or structure influences fire spread and development. Understanding the structure's floor plan, wall and ceiling penetrations, and interior stairwell placement will help you predict fire and smoke spread, locate the seat of the fire, and make accurate ventilation decisions. You will also be able to assign search and rescue activities based on this knowledge of the structure.

Many structures can have concealed vertical spaces that include pipe chases, HVAC ductwork, and vertical shafts that will allow fire spread to unaffected areas of the structure above the point of ignition. Concealed horizontal spaces are also a major concern. Examples of these spaces are continuous attic or cockloft spaces that connect individual retail spaces in strip malls where fire stopping is not required.

Figures 2.18a&b These photographs provide a visual comparison between (a) a balloon construction house and (b) a platform construction house. *(a) Courtesy of Wil Dane (b) Courtesy of Ron Moore.*

Figure 2.19 In some urban areas it is currently the trend to demolish older small houses and replace them with larger structures, sometimes referred to as McMansions.

Chapter 2 • Fire Behavior, Building Construction, and Building Occupancy

Voids or concealed spaces are present in almost every type of building construction. Their ability to allow a fire to grow and spread undetected is significant to a building's risk of destruction by fire more than any other building construction related factor. These spaces are of particular concern when the materials used to construct them are combustible. For example, a renovated building may have a new roof assembly over the existing roof(s) which may conceal HVAC components. This creates an additional undivided void space which is difficult to access. These roofs are also called rain or weather roofs because they are built over existing roofs.

Figure 2.20 Metal rain roofs are an inexpensive way to add protection to older flat roofs.

Rain roofs may be found on commercial buildings, schools, and residential structures **(Figure 2.20)**. Generally, they are placed over older flat roofs for aesthetic purposes, to prevent leaks, and to channel moisture off the roof. They may be constructed from lightweight metal panels and trusses or wood and roofing materials to form a peak or a flat roof surface. The void created by the rain roof can conceal a fire and allow it to burn undetected. As the trusses are exposed to fire, they will weaken, increasing the potential for collapse of both the rain roof and the original roof **(Figure 2.21)**. Ventilating a rain roof will not remove smoke from the structure until the original roof is penetrated. In addition, mechanical units placed under the rain roof place stresses on the existing roof and may cause the roof to collapse prematurely once exposed to fire from above or below.

Finally, knowledge of building construction is important to help you recognize the potential for structural collapse. Unprotected structural members that have been exposed to high temperatures from fire and the increased weight from fire suppression activities can collapse with little or no warning. Knowing the construction material and how fire affects the materials will assist in the risk benefit analysis. This information is gathered as part of the preincident survey and on-scene size-up for building construction.

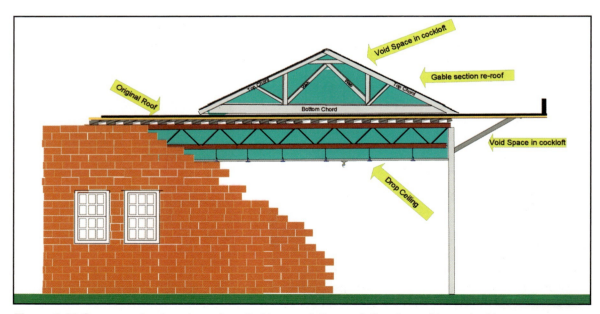

Figure 2.21 Components of a rain roof applied to an existing roof. *Courtesy of Larry Jenkins.*

Construction Type

The type of construction used in a structure is determined by the architect, structural engineer, or contractor. Locally adopted building codes are used to regulate the type of construction based on the occupancy, structure size, and the presence or absence of an automatic fire-suppression system. Building construction types are based upon the materials used in the construction as well as the fire-resistant characteristics of their structural components.

Building codes are adopted by the authority having jurisdiction (AHJ) and amended to meet local requirements. Some building codes are locally developed though most are based on nationally accepted model building codes. In the United States, there are currently two major model building codes, which are the NFPA® codes and the *International Building Code® (IBC®)* developed by the International Code Council (ICC). Canada has one building code that may be adopted by the provincial or local governments.

Not all buildings in your jurisdiction will be regulated by the local building code, however. Factory-built homes, sometimes referred to as mobile or manufactured homes, are generally exempt from local building codes. They are instead regulated by the U.S. federal government through the Department of Housing and Urban Development. It is also possible that federal or state buildings, including offices, courthouses, university buildings, and other government facilities, may be exempt from local building and fire code requirements **(Figure 2.22)**.

Figure 2.22 Federal and state/provincial buildings may not be required to meet municipal building and fire codes.

The following sections describe the general characteristics of each construction type specified in the model building codes for the U.S. and Canada. Although there are minor differences between the model building codes, generally these construction types are common to each. Because local AHJs can amend model codes to meet their needs, fire officers and firefighters must be familiar with the locally adopted building codes within their jurisdiction. In addition, information on factory-built homes is also included in this section.

Remember that when existing buildings are renovated, the result may be a structure containing more than one construction method. An example would be an ordinary construction commercial building originally built in the 1940s with a new addition of fire-resistive construction. In this case, operational decisions should be determined by the less fire-resistive construction method found in the original portion of the building.

NOTE: For more information about building construction, please refer to the IFSTA **Building Construction Related to the Fire Service** manual.

United States Construction

Both the *International Building Code® (IBC®)* and the National Fire Protection Association® (NFPA®) classify buildings in five types of construction (Type I through Type V). The types are then further divided into subcategories, depending on the code and construction type. Each construction type is defined by the construction materials and their performance under fire conditions. Every structure is composed of the following building elements:

- Structural frame
- Floor construction
- Roof construction

Besides the five classifications of construction, unclassified construction types also exist. They are comprised of the manufactured buildings that are completely assembled in a factory or on site from modules.

Type I (Fire Resistive)

Type I, fire resistive, construction provides the highest level of protection from fire development and spread as well as collapse **(Figure 2.23)**. All structural members are composed of only noncombustible or limited combustible materials with a high fire-resistive rating. Fire resistance is required to be 3 to 4 hours depending on the component, such as the wall, floor, ceiling, or roof. Type I construction can be expected to remain structurally stable during a fire and is considered to be most collapse resistant. Reinforced concrete and precast concrete along with **protected steel** frame construction meet the criteria for Type I construction.

Type I structures are often incorrectly referred to as being **fireproof**. Even though the structure will not burn, the structure may degrade from the effects of fire. Although the use of Type I construction provides structural stability should a fire occur, the addition of combustible materials in the form of contents, furniture, wall and window coverings, stock, and merchandise can generate sufficient heat over time to compromise the structural integrity of the building.

Type I construction buildings may contain a number of conditions that can affect their behavior during a fire, including the following:

- Compartments can retain heat, contributing to the potential for rapid fire development
- Unauthorized penetrations of fire barriers can permit the extension of fire into unaffected compartments
- Roofs may be difficult to ventilate due to construction material and design
- Windows may be solid causing them to be impossible to open for ventilation

> **Protected Steel** — Steel beams that are covered with either spray-on fire proofing (an insulating barrier) or fully encased in an Underwriters Laboratories Inc. (UL) designed system.
>
> **Fireproof** — Anything that is impervious to fire; also the act of protecting a structure or construction component from fire.

Figure 2.23 Typical Type I construction building. *Courtesy of Ron Moore.*

Type II (Noncombustible or Limited Combustible)

Buildings that are classified as Type II, noncombustible or limited combustible, construction are composed of materials that will not contribute to fire development or spread. Type II construction consists of noncombustible materials that do not meet the stricter requirements of those materials used in the Type I building classification **(Figure 2.24)**. Steel components used in Type II do not need to be protected or fire rated for the same lengths of time as Type I. Structures with metal framing members, metal cladding, or concrete-block construction of the walls with metal deck roofs supported by unprotected open-web steel joists are the most common form of this construction type. Fire-resistance rating is generally half that of Type I or 1 to 2 hours depending on the component. The lower fire-resistive rating of lighter weight materials used in this construction makes these buildings more prone to collapse.

Type II construction is normally used when fire risk is expected to be low or when fire suppression and detection systems are designed to meet the fuel load of the contents. You must always remember that the term *noncombustible* does not always reflect the true nature of the structure. Lower fire-resistive ratings are permitted for roof systems and flooring. Additionally, combustible features may be included on the exterior of Type II structures including balconies or facades for aesthetic purposes.

Type III (Ordinary Construction)

Type III ordinary construction is commonly found in older schools, mercantile buildings, and residential structures **(Figure 2.25)**. This construction type requires that exterior walls and structural members be constructed of noncombustible materials. Interior walls, columns, beams, floors, and roofs are completely or partially constructed of wood.

Type III construction buildings may contain a number of conditions that can affect their reaction during a fire, including the following:

- Voids exist inside the wooden channels created by roof and truss systems and between wall studs that will allow for the spread of a fire unless **fire stops** are installed in the void.

> **Fire Stop** — Solid materials, such as wood blocks, used to prevent or limit the vertical and horizontal spread of fire and the products of combustion in hollow walls or floors, above false ceilings, in penetrations for plumbing or electrical installations, in penetrations of a fire-rated assembly, or in cocklofts and crawl spaces.

Figure 2.24 Typical Type II construction building. *Courtesy of Ron Moore.*

Figure 2.25 An older Type III construction building that may not meet current building code requirements.

- Old existing Type III structures may have undergone renovations that have contributed to greater fire risk due to the creation of large hidden voids above ceilings and below floors which may create multiple concealed voids.

- New construction materials may have been removed to change the configuration or to open up the floor space during renovations. This may result in reducing the load-carrying capacity of the supporting structural member.

- The original use or occupancy of the structure may have changed to require a greater load-carrying capacity than the original design.

Type IV (Heavy Timber/Mill Construction)

Type IV, heavy timber, is characterized by the use of large-dimensioned lumber **(Figure 2.26)**. Dimensions vary depending on the particular building code being used; however, as a general rule these structural members will be greater than 8 inches (200 mm) in dimension with a fire-resistance rating of 2 hours. The dimensions of all structural elements, including columns, beams, joists, and girders must adhere to minimum dimension sizing. Any other materials used in construction and not composed of wood must have a fire-resistance rating of at least 1 hour.

Figure 2.26 Type IV construction consisting of heavy timber support beams. *Courtesy of Ron Moore.*

Type IV structures are extremely stable and resistant to collapse due to the sheer mass of their structural members. When involved in a fire, the heavy timber structural elements form an insulating shell derived from the timbers' own char that reduces heat penetration to the inside of the beam.

Exterior walls are constructed of noncombustible materials. Interior building elements such as floors, walls, and roofs are constructed of solid or laminated wood with no concealed spaces. This lack of voids or concealed spaces helps to prevent fire travel.

Modern Type IV construction materials may include small-dimensioned lumber that is glued together to form a laminated structural element. These elements are extremely strong and are commonly found in churches, barns, auditoriums, and other large facilities with vaulted or curved ceilings. (**Note:** When exposed to fire, these beams have the potential to delaminate as the glue is affected by heat and fail much in the same way as plywood.)

Type IV construction buildings may contain some conditions that can affect their behavior during a fire, including the following:

- The high concentration of wood can contribute to the intensity of a fire once it starts.

- Structural timbers exposed to fire can lose structural integrity causing attached masonry walls to collapse.

Type V (Wood Frame)

Type V construction is commonly known as wood frame or stick frame. The exterior **bearing walls** are composed entirely of wood. Occasionally, a veneer of brick or stone may be constructed over the wood framing **(Figure 2.27)**. The veneer offers the appearance of a Type III construction while providing little additional fire protection to the structure. Perhaps the most common example of this type of construction is a single-family residence.

> **Bearing Walls** — Walls of a building that, by design, carry at least some part of the structural load of the building in the direction of the ground or base.

Type V construction consists of framing materials that include wood 2 x 4 or 2 x 6 inch (50 mm by 100 mm or 50 mm by 150 mm) studs. Most structures built in northern climates mandate 6 inch (150 mm) exterior wall cavities for increased insulation. The outside of the framing members is covered with any one of a number of covering materials including:

- Aluminum siding
- Shake shingles
- Wood clapboards
- Sheet metal
- Cement
- Plastic (vinyl) siding
- Planks
- Plywood
- Composite wood (chipboard, particleboard, fiberboard)
- Styrofoam™ (which needs extra bracing) or stucco
- Veneers (brick or stone)

Figure 2.27 Example of a common Type V residential structure.

Exterior siding is attached by nails, screws, or glue. In the case of stucco, exterior siding is spread over a screen lattice that is attached to the framing studs.

Type V construction has evolved in recent years to include the use of prefabricated wood truss systems in place of the solid floor joist. The truss system creates a large, open void area between the floors of a structure rather than the closed channel system found with solid wood floor joists. When wood I-beams are used, they are usually constructed of thin plywood or wood composite, attached to beams that are 2 x 4 inch (50 mm by 100 mm) or smaller thickness, forming the top and bottom of the truss. These wood I-beams may have numerous holes cut in them to allow for electric, communication, and utility lines to be extended through them. Under fire conditions these plywood I-beams fail and burn much more rapidly than solid lumber.

Figure 2.28 Typical manufactured or mobile home found throughout the United States and in some parts of Canada.

A manufactured (mobile) home is one type of factory built structure that is popular in North America. *Manufactured homes* is a generic term used to describe structures that are completely built in a factory and shipped to the location where they are to be installed **(Figure 2.28)**. While manufactured homes take many forms, a characteristic of mobile homes is the existence of an axle assembly under the frame. Current estimates indicate that manufactured homes make up 25 percent of all housing sales in the U.S.

Manufactured homes are not required to conform to the model building codes although their construction type is similar to Type V construction. The age of the construction will make a difference in the fire resistance of the unit. Manufactured homes built before 1976 have less fire resistance than those of current construction. The major disadvantage of some of the manufactured homes is the fact that lightweight building materials used in them are susceptible to early failure in a fire. The heat produced by burning contents will cause the materials to ignite or melt rapidly. The contents have the same fuel loads as those found in conventional structures. At the same time, the use of lightweight materials makes forced entry much easier since walls can be quickly breached. In addition, some manufactured homes have open crawl spaces beneath them providing an additional source for oxygen during a fire.

NFPA® analysis of fires in residential occupancies indicates a steady decline in fires in manufactured homes since 1980. The analysis compares manufactured homes built before the Department of Housing and Urban Development (HUD) standard was enacted in 1976 (referred to as pre-standard) and those constructed after 1976 (post-standard). Reasons for the reduction in fire loss and fatalities can be contributed to requirements for:

- Factory-installed smoke alarms
- Use of flame retardant materials in interior finishes
- Use of flame-retardant materials around heating and cooking equipment
- Installation of safer heating and cooking equipment
- Installation of gypsum board rather than wood paneling in interior finishes

Unclassified Construction Types

Unclassified construction types include factory-built homes (also referred to as manufactured, prefabricated, and industrialized housing). Factory-built homes do not conform to the model building codes. However, their manufacture is regulated by the U.S. Department of Housing and Urban Development (HUD). Detailed requirements are included in the *Code of Federal Regulations — Title 24: Housing and Urban Development; Chapter XX, Part 3280*. Like the model building codes, this federal standard describes the fire-resistance requirements of materials used in the construction of these buildings.

There are five categories of factory-built homes:

- ***Manufactured:*** Of all the types of factory-built homes, manufactured homes are the most common type, almost completely prefabricated prior to delivery, and the least expensive. The HUD code preempts all local building codes and is more stringent than model building codes. Because the HUD code is based on performance standards, it tends to encourage construction innovations. Manufactured homes usually have a permanent steel undercarriage and are delivered on wheels towed by a transport vehicle. They normally range from one-section single-wide homes to three-section triple-wide homes.

- **Modular:** Modular or sectional homes must comply with the same local building codes as site-built homes. Only about 6 percent of all factory-built housing are modular homes. Modular sections can be stacked vertically and connected horizontally in a variety of ways. The modular section is transported to the site and then attached to a permanent foundation, which may include a full basement.

- **Panelized:** Panelized homes are assembled on-site from pre-constructed panels made of foam insulation sandwiched between sheets of plywood. The individual panels are normally 8 feet (2.44 m) wide by up to 40 feet (12.2 m) long. The bottom edges of the wall panels are recessed to fit over the foundation sill. Each panel includes wiring chases. Because the panels are self-supporting, framing members are unnecessary.

- **Pre-Cut:** Pre-cut homes come in a variety of styles including pole houses, post-and-beam construction, log homes, A-frames, and geodesic domes. The pre-cut home consists of individual parts that are custom cut and must be assembled on-site.

- **Hybrid modular:** One of the most recent developments in factory-built homes, the hybrid modular structure includes elements of both the modular design and the panelized design. Modular core units such as bathrooms or mechanical rooms are constructed in the factory, moved to the site and assembled. Pre-constructed panels are then added to the modules to complete the structure.

Canadian Construction

The *National Building Code of Canada (NBC)* defines the following three types of building construction:

1. **Combustible** — Construction that does not meet the requirements for noncombustible construction.

2. **Noncombustible** — Construction in which the degree of fire safety is attained by the use of noncombustible materials for structural members and other building assemblies.

3. **Heavy timber** — Combustible construction in which a degree of fire safety is attained by placing limitations on the sizes of wood structural methods and the thickness and composition of wood floors and roofs; also it avoids concealed spaces under floors and roofs.

To enable Canadian code users to understand these definitions, the NBC identifies specific requirements and limitations on materials used for each type of construction within the code. These requirements are listed in table formats that are easy to read and understand based on the occupancy classification and construction type.

In recent years, both the harsh winter climate and attempts at increasing energy efficiency have combined to increase the potential for structural collapse in new construction in Quebec, Canada. Residential dwellings may now be built under the design requirements of the *Novoclimat standard*. This standard is designed to make new homes more energy efficient and better insulated **(Figure 2.29, p. 70)**. The result is an almost airtight structure constructed with smaller dimensional lumber. Air is heated and circulated through a closed system of

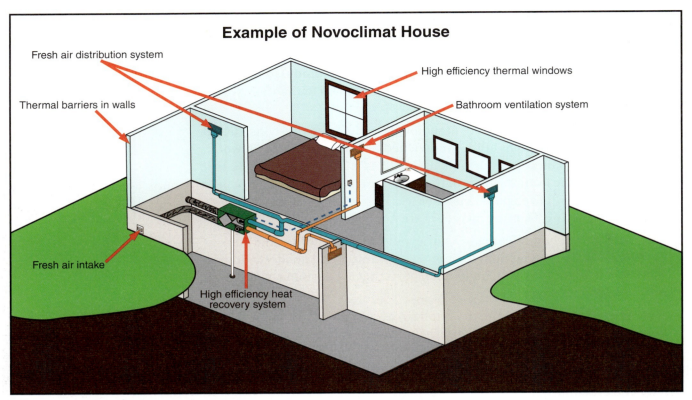

Figure 2.29 Typical energy efficient home design that is starting to appear in Canada known as the Novoclimat standard.

HVAC units, thermo-pumps, and other devices. The result for firefighters is a structure more likely to fail rapidly under fire conditions as thermal heat and fire gases are contained within the compartments and structure. The design also makes vertical ventilation more difficult.

Building Periods or Age

Knowing how old a structure is can help you determine how fire will behave in the structure. In North America, the oldest structures still in use are found in Florida, the Northeastern states, and the Canadian provinces. Urban centers such as New York City, Boston, Philadelphia, St. Augustine, Montreal, and Quebec City contain structures that have existed before the 1800s. Most cities, however, contain structures built during the Industrial Revolution of the late 1800s when the use of steel structural members became prevalent. At the same time, structures grew in height with the advent of elevators and the use of steel frames in high rise buildings.

Early Regulatory Codes

With the increase in urban development in the late 1800s, municipalities began to regulate the construction of new structures. As cities developed their own building codes, the need for uniformity of construction and increased fire protection became evident. Over the past century, the application of building code requirements has become universal in the majority of jurisdictions in North America.

The economic boom following World War II generated two trends in the construction industry. First was the increase in high rise structures in both new urban areas and existing cities. Urban renewal projects removed deteriorating structures to make way for new structures. New advances in HVAC systems resulted in the use of large expanses of open space and return air plenums above drop ceilings. HVAC systems made natural ventilation unnecessary and caused the addition of nonfunctioning glass panels on the exterior of structures that limited access from the exterior. The HVAC systems also created new avenues of fire spread within the enclosed structure.

Suburban Development

The second trend was in the construction of single family dwellings in suburban areas surrounding the metropolitan centers **(Figure 2.30)**. The rapid expansion of the suburbs and the need for inexpensive housing resulted in a shift from balloon construction to platform construction. Platform construction required shorter pieces of lumber resulting in lower materials costs. At the same time, fire stops were added to the wall cavities to prevent fire spread, creating a safer structure.

Figure 2.30 The Post-World War II era saw the spread of the suburbs typified by similar style houses in vast tracks like the one pictured in Pennsylvania. *Courtesy of Library of Congress.*

Along with single family dwellings, apartment buildings spread into the suburban areas changing in appearance and arrangement. Apartment designs shifted from rectangular brick buildings with a front and rear entrance, internal hallways, and 4-stories in height to wood frame buildings with external access to each unit, continuous cocklofts, and heights of 2 to 4 stories. Fire codes began to require 1-2 hour rated fire walls between individual units. In an effort to limit fire spread, to limit the area of common cocklofts, and to separate structures into areas of no more than 4 units, fire walls that were continuous from foundation to above the roof line were required.

New Material Uses

Changes in the types of building materials used has also added to potential fire hazards. During the 1960s, the use of Romex® wiring containing one strand of copper and one of aluminum became prevalent. Aluminum was used due to a shortage of copper which was in high demand for the manufacture of ammunition needed in the Vietnam War. The use of two dissimilar metals in wiring resulted in electrical fires in receptacles and switch boxes. The dissimilar metals expanded at different rates when heated and caused connections to loosen which then caused arcing and fires. This continues to be a problem as these structures age.

From the mid-1980s to the mid-1990s almost one million residential roof structures were constructed utilizing fire retardant plywood sheathing. Unfortunately, many of these roofs began disintegrating at a rapid speed due to the heat buildup in attics from solar radiation and the moisture content of the wood. The average life span of these roofs has proven to be less than 10 years, with the majority lasting 3 to 8 years. Although most of these roofs have been replaced, some may still be found.

If you know the time period in which your immediate response area was developed, you will be able to determine the general construction types and building code requirements. It is important to remember that in older communities, some structures may not have been built to any code requirements while additions made to them were required to meet the codes in effect at the time of the addition. You should consult your fire prevention division or building department to determine the requirements for some of the buildings that you are responsible for protecting.

Interior Building Arrangement

Most buildings are simply boxes that are designed to contain different types of activities or functions. With the exception of common residential structures, the actual interior arrangement of most buildings can be difficult to determine from the outside. Even visual clues like the location and arrangement of doors, windows, chimneys, wall height, and the number of stories cannot provide you with the exact indication of the internal floor plan. However, it does provide a general idea of the room arrangement.

The interior arrangement of a structure influences fire behavior within the structure. For the purposes of this manual, interior arrangement of a structure can be classified as *open floor plan* or *compartmentalized*.

Open Floor Plans

An open floor plan arrangement lacks interior floor-to-ceiling walls to break up the area into smaller compartments. Many of the exterior walls are generally visible from any point inside the open floor plan. The spaces created by the open floor plan may have high, multistory ceilings, such as the Big Box retail stores, or ceilings that are 8 to 10 feet (2.44 to 3.05 m) such as those found in office spaces on a single floor of a high rise structure. The open floor plan can be found in warehouses, offices, auditoriums, and houses of worship among others. Open floor plan offices may be subdivided into individual work spaces by partitions that are 4 to 6 feet (1.2 to 1.83 m) high. The partitions, which create cubicles, generally do not reach the ceiling and may not be secured to the floor or a permanent wall assembly.

In single-family residences built using the open floor plan design, there will be no walls between the kitchen, family, dining, and living rooms. High ceilings are a common design feature of new construction residences of this type **(Figure 2.31)**. Walls will usually be provided for bathrooms, bedrooms, utility rooms, and storage spaces. While the interior walls of residences are not always fire resistant, the wall between the garage and the dwelling is required to be fire resistant by building codes.

Fires that begin in an open floor plan structure will be fuel controlled because they will have adequate oxygen available. Ceiling height will influence the increase in temperature at the ceiling level. Fire spread can be influenced by air currents within the space caused by exterior openings or HVAC systems. The lack of fire barrier walls will permit the fire to spread by conduction

Figure 2.31 Open floor plan houses have common areas for living, dining, and cooking.

from one fuel item to the next. If the fire is allowed to spread undetected or uninterrupted, the interior temperature will increase to the point where it can weaken the roof structure and cause it to collapse.

Compartmentalization

A compartmentalized structure contains interior walls that create small boxes or spaces within the confines of the exterior walls and roof or ceiling. Compartmentalized layouts are typical for residential-type structures built before the 1980s where the structure is divided into multiple rooms or compartments. This arrangement is also common in commercial structures, office buildings, schools, and institutions **(Figure 2.32)**.

Figure 2.32 Compartmentalized offices have corridors connecting work areas and entrance lobbies along with other function areas.

A fire in a compartmentalized structure will generally be ventilation-controlled based on the amount of oxygen that is available within the compartment. Fire spread will be limited by the fire-resistive quality of the compartment's walls, ceiling, doors, and floor. If the compartment contains sufficient fuel and oxygen, however, the fire can reach a magnitude that will breach even fire resistive barriers and extend into other compartments.

Basements, Cellar, and Crawl Spaces

While many buildings are built on slab foundations, some new buildings and many older ones are built with a space between the first floor and the soil. This space may be large enough for human occupancy, such as a basement or cellar, or simply a space for ventilation, such as a crawlspace **(Figures 2.33a-c)**.

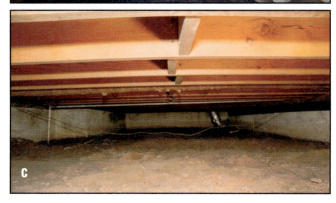

Figures 2.33a-c There are three common areas that may be found under a residential structure: (a) finished basement; (b) cellar; and (c) crawl space.

Depending upon the geographic region, the terms *basement* and *cellar* are interchangeable. Both are used to describe a room or space beneath a structure wholly or partially below ground level that may or may not be used for a living space. The space may be unfinished, with the foundation forming the exterior walls and the first floor joists visible, or covered with an interior finish such as Sheetrock®. The floor may be dirt, concrete, or stone. Basements may contain the structure's HVAC system including fuel storage such as coal or fuel oil. Cellars are generally smaller unheated spaces used to store food or wine at cooler ambient temperatures. Cellars may also be designed to provide protection from tornadoes or other storms. In either case, access may be gained via interior stairs or an exterior entrance.

Crawl spaces are found in single family dwellings and small commercial structures. Crawl spaces are formed by a perimeter wall foundation with a subfloor that is framed with wood beams and floor joists. In recent years, this type of construction has been replaced by concrete slab construction. Crawl spaces were originally used to provide ventilation for the structure allowing cool air to circulate beneath the ground floor and separate the floor from the ground below. Floor furnaces, fueled by natural gas, may be found in crawlspaces, depending on the crawlspace for oxygen for the unit's combustion chamber.

Many firefighters believe that basement and cellar fires are the hardest types of structure fires to control. Conditions that make these fires hard to control include:

- Most basements have only one entry point that will act like a chimney during a fire by funneling heat, fire, and smoke into the path of fire crews.
- The ceiling above the basement is generally exposed and the ceiling height is approximately 8 feet (2.44 m) high, allowing the fire to rapidly spread upward igniting the first floor joists.
- Basements under balloon construction buildings permit the fire to spread up the insides of exterior walls as far as the roof.
- Fires in crawl spaces are difficult to contain due to the lack of access to the area.
- Ventilation options are limited in basement/cellar and crawlspace fires, contributing to extinguishment difficulties.

Basement Fire in Washington D.C. Townhouse Results in Two LODDs

On May 30, 1999, two firefighters died and two were injured as the result of a fire in the basement of a multistory brick townhouse in Washington, D.C. The casualties were located on the first floor of the structure. The firefighters had entered through the front door believing that the fire was located on that level. An engine and a ladder responding to the back of the structure reported seeing the fire when they opened the grade level door. This door actually led into the basement and not the first floor as the IC believed. Permission to attack the fire was not given by the IC who did not want opposing fire streams. Opening the basement door resulted in a rapid increase in the magnitude of the fire, forcing it up the interior stairs and onto the crews operating on the first floor. During a rapid withdrawal by the first floor crews, two firefighters became lost while others exited with burns. When it was determined that two men were missing, rescue personnel reentered the building and located one by his activated PASS device. The other was located 4 minutes later; his PASS device had not been turned on and did not activate.

NIOSH fire fighter fatality report (FACE#99F21) contains a number of recommendations that could have prevented these deaths. Among these are:

- Ensure that the department's Standard Operating Procedures (SOPs) are followed and refresher training is provided
- Provide the Incident Commander with a Command Aide
- Ensure that firefighters from the ventilation crew and the attack crew coordinate their efforts
- Ensure that when a piece of equipment is taken out of service, appropriate backup equipment is identified and readily available
- Ensure that personnel equipped with a radio position the radio to receive and respond to radio transmissions
- Consider using a radio communication system that is equipped with an emergency signal button, is reliable, and does not produce interference
- Ensure that all companies responding are aware of any follow-up reports from dispatch
- Ensure that a Rapid Intervention Team is established and in position immediately upon arrival
- Ensure that any hoseline taken into the structure remains inside until all crews have exited
- Consider providing all firefighters with a Personal Alert Safety System (PASS) integrated into their Self-Contained Breathing Apparatus (SCBA)
- Develop and implement a preventive maintenance program to ensure that all SCBAs are adequately maintained

Attics and Cocklofts

The space between the top floor of a structure and the roof is referred to as the *attic* or *cockloft* **(Figures 2.34a&b, p. 76)**. While both terms refer to similar spaces, an attic is usually found in residential structures and is large enough for a person to walk upright. Attics may be finished, with flooring and interior finish on the underside of the roof, or they may be unfinished with the roof rafters and ceiling joists exposed. They are generally used for storage but may also contain an HVAC unit to supply the upper floor of the structure. A cockloft is a space 2 to 3 feet (.06 to 0.9 m) in height that is found over commercial building spaces and is not designed for human habitation. It is often found beneath flat roofs and may be accessed through a hatch from below. In strip malls, the cockloft may extend throughout the structure connecting individual occupancies.

Attic and cockloft fires are very difficult to access and control. They also make vertical ventilation difficult by creating an unsafe condition for firefighters working on the roof. Most strip mall buildings today have truss roof assemblies that create an extremely dangerous condition when exposed to fire. In strip malls, fires may extend across several occupancies and spread

Figures 2.34a&b The space between the uppermost ceiling and the roof may be referred to as (a) an attic or (b) as a cockloft.

rapidly without the original point of origin being evident. Locating cockloft fires requires both opening the roof and pulling ceilings down to determine the extent of fire spread. In occupancies that have high ceilings, this activity may require the use of long pike poles, ground ladders, or a combination. The addition of drop ceilings installed under existing ceilings and rain roofs over existing roofs can make cockloft fires extremely difficult to locate and extinguish. Controlling the attic in structures that share a common attic is key to controlling fire spread and fire extinguishment.

Concealed Spaces

The term *concealed space* (*void space*) is used to describe any area between wall surfaces, over ceilings, and under floors that are not visible from the normal occupied area. Concealed spaces are generally designed to provide passive insulation and sound barriers. They may also provide paths for return air from the compartments to the HVAC system called **plenums**, kitchen exhaust ducts, or as spaces for electrical and plumbing systems to connect multiple floors of a structure. Trash and laundry chutes may also exist in concealed spaces. A concealed space is not visible unless an access panel is provided to it.

> **Plenum** — In a structure, the space between the real ceiling and a suspended ceiling. It is often used as a return air duct for the heating and air conditioning system. It may also contain electrical, telephone, and communication wires.

Concealed spaces create pathways for fire spread from one floor to the next or from compartment to compartment **(Figure 2.35)**. Although rare, fires may originate in a concealed space caused by an electrical short circuit in the wiring or a leak in a natural gas line within the space. Controlling a fire spreading through a concealed space requires gaining access to the space to apply water directly to the fire.

Structural Collapse Potential

The structural failure of a building or any portion of it resulting from a fire, snow, wind, water, or damage from other forces is referred to as **structural collapse**. Structural collapse can also be the result of an explosion, earthquake, flood, or other natural occurrence. However, natural or explosion caused collapses usually occur without warning prior to an emergency response while fire caused collapses often occur during emergency operations. The ability to understand how fire can cause building elements to deteriorate resulting in a collapse is extremely important for fire officers and firefighters. Collapse

> **Structural Collapse** — Structural failure of a building or any portion of it resulting from a fire, snow, wind, water, or damage from other forces.

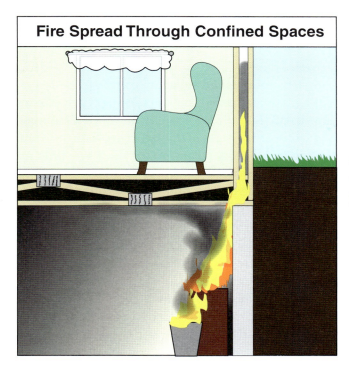

Figure 2.35 Large or small voids can contribute to the spread of fire between floors.

potential should be considered during preincident surveys and throughout the size-up process until the situation is mitigated and control of the property is returned to the owner or other authorities.

Vendome Hotel Fire
Boston, Massachusetts, June 17, 1972

At 2:35 p.m. on Saturday, June 17, 1972, the first of four alarms was received for a fire in the former Hotel Vendome. It took nearly three hours to stop the blaze. When the fire was extinguished, firefighters began overhaul operations at the site. At 5:28 p.m., without warning, the southeast section of the building collapsed killing nine firefighters and injuring eight others. The firefighters, along with one ladder apparatus, were buried under a two-story pile of debris.

Although the cause of the fire was not determined, the cause of the collapse was attributed to renovations made to the structure in 1971. During the renovation, a seven inch (178 mm) steel support column was cut to make room for a new HVAC duct. This weakened the column causing it to collapse under the added weight of water used to control the fire on the upper floors.

Factors that should be considered when trying to determine the potential for structural collapse are:

- Construction type
- Age of structure
- Renovations, additions, and alterations
- Contents
- Length of time the fire has been burning

- Stage of the fire
- Amount of water used to extinguish the fire
- Weather

A collapse or safety zone must be established adjacent to any exposed exterior walls of the structure. Apparatus and personnel operating master stream appliances must not be positioned in the collapse zone. Traditionally, **collapse zones** have been estimated by taking the height of the structure and multiplying it by a factor of 1½. For example, a 3-story structure that is 30 feet (9.14 m) tall would require a collapse zone no less than 45 feet (13.72 m) from the base of the structure. As structures increase in height, this becomes impractical by creating a space that limits defensive fire fighting operations. Therefore, the following guidelines should be considered when determining a collapse zone:

- **Type I** construction high rise buildings are not as likely to collapse, making the primary concern the hazard of flying glass from windows or curtain walls. In Type I construction, it is the contents of the building burning, not the structure itself. Collapse zones must be determined considering the direction and velocity of wind currents that can carry the glass shards. Structural collapse, if it does occur, will be localized and not structure wide.

- **Type II** construction consists of unprotected steel or noncombustible supports, such as I-beams. When exposed to temperatures above 1,000°F (538°C), unprotected steel will expand and twist, pushing out walls, and when cooled will slightly constrict. These movements will cause floors and walls to collapse. Any type of construction that includes brick and block walls supporting unprotected steel bar joists and I-beams are involved in a large number of these collapses.

- **Type III** construction multistory buildings should have a collapse zone of 1½ times the height of the structure. For example, a building 7 stories (70 feet [21.34 m]) will require a collapse zone of approximately 105 feet (32 m) during defensive operations. In Type III ordinary construction, exterior load bearing walls are made of concrete, brick, or masonry while interior loads are carried by wood, masonry, or unprotected steel. Masonry construction walls can collapse in one piece or crumble in many parts. When the debris strikes the ground it can travel a distance and even cause the collapse of other structures or objects.

- **Type IV** heavy timber or mill construction is one of the least likely to collapse. The weight-bearing capacity of the large dimension wood members will resist collapse unless they have been affected by a large volume of fire. A collapse zone should be established if the fire is intense or the structure has been weakened by repeated fires over time.

- **Type V** construction collapses are influenced by the style of construction. That is, a multistory platform structure will generally burn through and collapse inward while a balloon structure can have full walls fall outward in a single piece. Exterior masonry and veneer walls that are not load bearing are placed over load bearing wood walls. Brick veneer attached to the frame can fall straight down (*curtain collapse*) into a pile or fall as a unit straight out as the ties and supports fail. Although it is rare for a Type V building to collapse outward, there is a great danger to firefighters due to interior collapses. Lightweight trusses will fail within five minutes when exposed to direct heat.

> **Collapse Zone** — The area extending horizontally from the base of the wall to one and one-half times the height of the wall.

In North America, examples of structural collapse involving high rise buildings or Type I construction buildings is very limited. Strict building codes have ensured that structural members exposed to fire and high temperatures will remain sound until the fire is extinguished. Structural collapses due to earthquakes generally involve smaller buildings such as the one- to four-story buildings in the Marina Section of San Francisco in 1989. Some of these buildings simply fell over or against another building.

Collapse zones should be established when there is an indication that the structure has been weakened by prolonged exposure to fire or heat, when a defensive strategy has been adopted, or when interior operations cannot be justified. The size of the collapse zone must take into consideration the type of building construction, other exposures, and the safest location for apparatus and personnel. Church steeples, water tanks, chimneys, and false facades that extend above the top of the structure must be viewed as a potential collapse hazard even if the structure is not. Most collapses usually involve brick or masonry block and may be structural components or veneer. Structural collapses are not limited to the actual emergency and can occur well after the fire is extinguished. Fire inspectors must ensure the structural stability of the site before entering it.

Because the collapse zone extends the full length of all of the affected walls, the safest location for defensive operations is at the corner of the building **(Figure 2.36)**. Master streams and apparatus can be located in the area formed by a 90 degree arc from the wall intersection as long as they are far enough away that flying debris will not strike them.

The longer a fire burns, the greater the temperature of the fire gasses in the upper levels of the structure or compartment. In 1918, the American Society for Testing and Materials (ASTM) developed the standard time-temperature curve that is still used to illustrate the rate of temperature increase in a compartment fire **(Figure 2.37, p. 80)**. By applying an estimate of the length of time the fire has been burning to the type of construction, you will be able to get a general idea of how hot the ceiling and roof components have gotten.

The stage of the fire can easily indicate the quantity of heat that the structure has been exposed to and the potential for structural collapse. Fires in the incipient stage will not have generated sufficient heat or flame to cause unprotected steel or wood frame construction to collapse. However, collapse potential increases in the growth stage as heat increases in the upper levels of the space and flame spreads to and consumes the combustible structural members. In the decay stage, and during post-suppression activities, collapse becomes very likely due to the weakened state of structural members and the buildup of water.

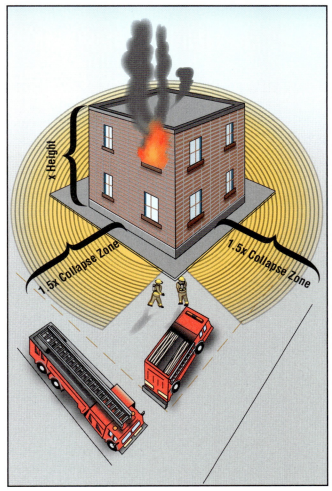

Figure 2.36 The safest position for access to many structures without being in the collapse zone is in the angle formed by the corners.

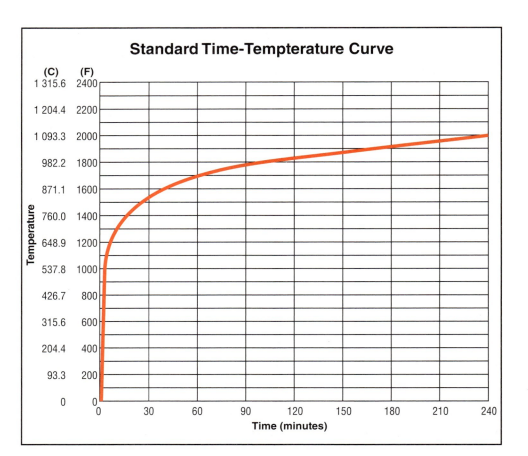

Figure 2.37 Example of the time-temperature curve.

Another factor that can contribute to structural collapse is the contents within the structure or on the roof. The contents may contribute to collapse in three ways: First, by adding to the fuel load in the building and generating higher temperatures and rapid combustion that will weaken the structure. Second, by adding weight to the weakened structural members causing them to collapse more rapidly. Finally, the ability of the contents to retain water increases their weight and the stress on the structural members. Contents include such things as stored materials, furniture, and machinery. Like knowledge of the construction type, knowledge of the contents is gained through preincident surveys and inspections.

While contents within a structure are visible during preincident surveys, storage in concealed spaces and attics of residences is not. Attic storage is often heavier than the ceiling joists have been designed to carry. When the joists are weakened by fire, storage increases the potential for ceiling joists to fail, putting firefighters at risk. Such storage is common in residential dwellings and has been the cause of firefighter fatalities in industrial fires.

Finally, the quantity of water that is used to suppress the fire can have a direct effect on an unstable structure. Every US gallon (SI liter) of water that is used to suppress the fire adds 8.33 pounds (3.78 kilograms) (Imperial gallon 10 pounds) of weight to floors that may already be weakened. The added weight may cause floors to pancake down or push walls out resulting in a complete failure of the structure. As an estimate, 250 gpm (1 000 L/min) adds 1 ton (900 kg) of water per minute to the structure.

In addition to the factors previously listed, indicators of potential or imminent collapse include:

- Roof sagging, pulling away from parapet walls, or feeling spongy (soft) underfoot
- Floors sagging or feeling spongy (soft) underfoot
- Chunks of ceiling tiles or plaster falling from above
- Movement in the roof, walls, or floors
- Noises caused by structural movement
- Little or no water runoff from the interior of the structure
- Cracks appearing in exterior walls with smoke or water appearing through the cracks
- Evidence of existing structural instability such as the presence of tie rods and stars that hold walls together (**Figure 2.38**)
- Loose bricks, blocks, or stones falling from buildings
- Deteriorated mortar between the masonry
- Walls that appear to be leaning
- Structural members that appear to be distorted
- Fires beneath floors that support heavy machinery or other extreme weight loads
- Prolonged fire exposure to the structural members (especially trusses)
- Structural members pulling away from walls
- Excessive weight of building contents

Figure 2.38 Symbolic of reinforced walls in older structures are cast iron stars on exterior walls.

WARNING! Structural collapse can occur with little warning. If indicators start to appear, collapse is imminent and personnel must be withdrawn from the structure and the collapse zone.

Until recently there has not been any documented evidence of how long structural members will remain intact when exposed to fire. The National Institute of Standards and Technology (NIST) and the United States Fire Administration (USFA) began testing in 2000 to determine how firefighters could predict potential structural collapse. Although the analysis is still ongoing, some of the results have been released. Tests in both residential and commercial structures indicate that:

- Steel bar joist supported roofs will collapse 10 to 20 minutes after ignition.
- Wood truss supported roofs will collapse 15 to 20 minutes after ignition.

Collapse of structures using lightweight construction can occur earlier in the incident and may not provide the warning indicators listed above. A thorough preincident survey and size-up of the incident scene will provide you with some indication of the presence of lightweight construction.

Safety Alert

Firefighters may be injured or killed when fire-damaged roof and floor truss systems collapse, sometimes without warning. Firefighters should take the following steps to minimize the risk of injury or death during structural firefighting operations involving roof and floor truss systems:

- Know how to identify roof and floor truss construction.
- Report immediatley the presence of truss construction and fire involvement to the Incident Commander (IC).
- Use a thermal imaging camera as part of the size-up process to help locate fires in concealed spaces.
- Use extreme caution and follow standard operating procedures (SOPs) when operating on or under truss systems.
- Open ceilings and other concealed spaces immediately whenever a fire is suspected of being in a truss system. *Other guidelines to follow:*
  — Use extreme caution because opening concealed spaces can result in backdraft conditions.
  — Always have a charged hoseline available.
  — Position between the nearest exit and the concealed space to be opened.
  — Be aware of the location of other firefighters in the area.
- Understand that fire ratings may not be truly representative of real-time fire conditions and the performance of truss systems may be affected by fire severity.
- Take the following steps to protect firefighters before emergency incidents:
  — Conduct preincident planning and inspections to identify structures that contain truss construction.
  — Ensure that firefighters are trained to identify roof and floor truss systems and use extreme caution when operating on or under truss systems.
  — Develop and implement SOPs to safely combat fires in buildings with truss construction.
- Use the following procedures to protect firefighters at the emergency incident:
  — Ensure that the IC conducts an initial size-up and risk assessment of the incident scene before beginning interior firefighting operations.
  — Evacuate firefighters performing operations under or above trusses as soon as it is determined that the trusses are exposed to fire, and move to a defensive mode.
  — Use defensive overhauling procedures after extinguishing a fire in a building containing truss construction.
  — Use outside master streams to soak smoldering trusses and prevent rekindles.
  — Report any damaged sagging floors or roofs to Command.

Source: National Institute for Occupational Safety and Health (NIOSH): "Preventing Injuries and Deaths of Fire Fighters Due to Truss System Failures," NIOSH Publication No. 2005-123.

Occupancy Types

Occupancy types are the classifications of the use a structure is designed to contain. Occupancy classifications are defined by the building code and life safety code adopted by the local jurisdiction. The three primary model building codes in use in North America are NFPA®, IBC®, and the National Building Code of Canada (NBC). **Table 2.1, p. 84-87,** provides a comparison of the occupancy classifications found in these three codes

Structures may be divided into either single-use occupancies or multiple use occupancies. These are described in the following sections.

Single Use

A structure that is designed for a single use must meet the building code requirements for that use. For instance, an office building must meet the requirements found in the Business Occupancy Classification while an elementary school must meet the requirements of an Educational Occupancy. Requirements include exit access, emergency lighting, fire protection systems, construction type, and fire separation barriers among many others.

In many cases, a structure such as an industrial facility will contain multiple types of uses including storage, processing, and office. If the structure has one owner and occupant, then the structure is generally classified by its primary function. For example, a restaurant located in a single structure would be classified as an Assembly Occupancy even though it also contains areas for cooking, washing, and multiple types of storage (**Figure 2.39**). Fire separations and fire protection systems may be required to protect one area from another.

Multiple Use

Structures that contain multiple-use occupancies must meet the requirements for each individual occupancy classification. That is, in a strip mall, each space is classified by its use and separated from the other units by a fire-rated assembly or wall, as required by the building code. Within the strip shopping mall, there may be retail outlets [Mercantile Occupancy], offices [Business Occupancy], and small restaurants [Assembly Occupancy] (**Figure 2.40**). While the concept that each space will be separated from the others and meet its own require-

Figure 2.39 A freestanding restaurant is an example of a single-occupancy structure found in most communities.

Figure 2.40 A strip mall may be a multi-occupancy structure depending on the types of businesses located in the various units.

Table 2.1

This table is a general comparative overview of the occupancy categories for three major model code systems. Readers must consult the locally adopted code and amendments for complete information regarding each of these occupancies.

Occupancy	IBC®	NFPA®	NBC
Assembly	**A-1** - occupancies with fixed seating that are intended for the production and viewing of performing arts or motion picture films. **A-2** - those that include the serving of food and beverages; occupancies have nonfixed seating. Nonfixed seating is not attached to the structure and can be rearranged as needed. **A-3** - occupancies used for worship, recreation, or amusement, such as churches, art galleries, bowling alleys, amusement arcades, as well as those that are not classified elsewhere in this section. **A-4** - occupancies used for viewing of indoor sporting events and other activities that have spectator seating. **A-5** - outdoor viewing areas; these are typically open air venues but may also contain covered canopy areas as well as interior concourses that provide locations for vendors and other commercial kiosks.	**Assembly Occupancy** - An occupancy (1) used for a gathering of 50 or more persons for deliberation, worship, entertainment, eating, drinking, amusement, awaiting transportation, or similar uses; or (2) used as a special amusement building, regardless of occupant load.	**Group A Division 1** - Occupancies intended for the production and viewing of the performing arts. **Group A Division 2** - Occupancies not classified elsewhere in Group A. **Group A Division 3** - Occupancies of the arena type. **Group A Division 4** - Occupancies in which occupants are gathered in open air.
Business	**Business Group B** - Buildings used as offices to deliver service-type or professional transactions, including the storage of records and accounts. Characterized by office configurations to include: desks, conference rooms, cubicles, laboratory benches, computer/data terminals, filing cabinets, and educational occupancies above the 12th grade.	**Business** - Occupancy used for the transaction of business other than mercantile.	**Group D** - Business and personal services occupancies

Table 2.1
Continued

Occupancy	IBC®	NFPA®	NBC
Educational	**Educational Group E** - Buildings providing facilities for six or more persons at one time for educational purposes in grades kindergarten through twelfth grade. Religious educational rooms and auditoriums that are part of a place of worship, which have occupant loads of less than 100 persons, retain a classification of Group A-3.	**Educational Occupancy** - Occupancy used for educational purposes through the twelfth grade by six or more persons for 4 or more hours per day or more than 12 hours per week.	Covered under Group A
Factory Industrial	**Factory Industrial Group F** - Occupancies used for assembling, disassembling, fabrication, finishing, manufacturing, packaging, repair, or processing operations. - **Factory Industrial F-1 Moderate Hazard** (examples include but not limited to: aircraft, furniture, metals, and millwork) - **Factory Industrial F-2 Low Hazard** (examples include but not limited to: brick and masonry, foundries, glass products, and gypsum) **High Hazard Group H** - Buildings used in manufacturing or storage of materials that constitute a physical or health hazard. - **High-Hazard Group H-1** - detonation hazard - **High-Hazard Group H-2** - deflagration or accelerated burning hazard - **High-Hazard Group H-3** - materials that readily support combustion or pose a physical hazard - **High-Hazard Group H-4** - health hazards - **High-Hazard Group H-5** - hazardous production	**Industrial Occupancy** - Occupancy in which products are manufactured or in which processing, assembling, mixing, packaging, finishing, decorating, or repair operations are conducted.	**Group F Division 1** - High-hazard industrial occupancies **Group F Division 2** - Medium-hazard occupancies **Group F Division 3** - Low-hazard industrial occupancies

Table 2.1
Continued

Occupancy	IBC®	NFPA®	NBC
Occupancy Institutional (Care and Detention)	**Institutional Group I** **Group I-1 -** Assisted living facilities holding more than 16 persons on a 24-hour basis. These persons are capable of self-rescue. **Group I-2 -** Medical, surgical, psychiatric, or nursing care facilities for more than five people who are not capable of self-preservation or need assistance to evacuate. **Group I-3 -** Prisons and detention facilities for more than five people under restraint. **Group I-4 -** Child and adult day care facilities.	**Ambulatory Health Care -** Building (or portion thereof) used to provide outpatient services or treatment simultaneously to four or more patients that renders the patients incapable of taking action for self-preservation under emergency conditions without the assistance of others. **Health Care -** An occupancy used for purposes of medical or other treatment or care of four or more persons where such occupants are mostly incapable of self-preservation due to age, physical or mental disability, or because of security measures not under the occupants' control. **Residential Board and Care -** Building or portion thereof that is used for lodging and boarding of four or more residents, not related by blood or marriage to the owners or operators, for the purpose of providing personal care services. **Detention and Correctional -** An occupancy used to house one or more persons under varied degrees of restraint or security where such occupants are mostly incapable of self-preservation because of security measures not under the occupants' control.	**Group B Division 1 -** Care or detention occupancies in which persons are under restraint or are incapable of self-preservation because of security measures not under their control. **Group B Division 2 -** Care or detention occupancies in which persons having cognitive or physical limitations require special care or treatment.
Mercantile	**Mercantile Group M -** Occupancies open to the public that are used to store, display, and sell merchandise with incidental inventory storage.	**Mercantile -** An occupancy used for the display and sale of merchandise.	**Group E -** Mercantile occupancies
Residential	**Residential Group R** **R-1 -** Residential occupancies containing sleeping units where the occupants are primarily transient in nature (boarding houses, hotels, and motels) **R-2 -** Residential occupancies containing sleeping units or more than 2 dwelling units where the occupants are primarily permanent in nature (apartments, convents, non-transient hotels, etc...)	**Residential Occupancy -** Provides sleeping accommodations for purposes other than health care or detention and correctional. **One- and Two-Family Dwelling Unit -** Building that contains not more than two dwelling units with independent cooking and bathroom facilities.	**Group C -** Residential occupancies

Table 2.1
Continued

Occupancy	IBC®	NFPA®	NBC
Residential (continued)	**R-3** - Residential occupancies where the occupants are primarily permanent in nature and not classified as Group R-1, R-2, R-4, or I **R-4** - Residential occupancies shall include occupancies buildings arranged for occupancy as residential care/assisted living facilities for more than 5 but less than 16 occupants (excluding staff)	**Lodging or Rooming House** - Building (or portion thereof) that does not qualify as a one- or two-family dwelling, that provides sleeping accommodations for a total of 16 or fewer people on a transient or permanent basis, without personal care services, with or without meals, but without separate cooking facilities for individual occupants. **Hotel** - Building or groups of buildings under the same management in which there are sleeping accommodations for more than 16 persons and primarily used by transients for lodging with or without meals. **Dormitory** - A building or a space in a building in which group sleeping accommodations are provided for more than 16 persons who are not members of the same family in one room, or a series of closely associated rooms, under joint occupancy and single management, with or without meals, but without individual cooking facilities. **Apartment Building** - Building (or portion thereof) containing three or more dwelling units with independent cooking and bathroom facilities.	**Group C** - Residential occupancies
Storage	**Storage Group S** – Structures or portions of structures that are used for storage and are not classified as hazardous occupancies. - **Moderate-Hazard Storage, Group S-1** (examples include but not limited to: bags, books, linoleum, and lumber) - **Low-Hazard Storage, Group S-2** (examples include but not limited to: asbestos, bagged cement, electric motors, glass, and metal parts)	**Storage Occupancy** - An occupancy used primarily for the storage or sheltering of goods, merchandise, products, vehicles, or animals.	Covered under Group F
Utility/ Miscellaneous	**Utility/Miscellaneous Group U** - These are accessory buildings and other miscellaneous structures that are not classified in any specific occupancy (agricultural facilities such as barns, sheds, and fences over 6 ft [2 m])	—	—

Size-Up: Evaluation and Assessment

Chapter Contents

Size-Up Application 93
    Preincident.. 94
    While Responding 102
    Facts, Preceptions, and Projections..........104
    On Arrival 106
    During the Incident 111

Incident Size-Up Considerations 111
    Facts 112
    Probabilities 113
    Own Situation 114
    Decision-Making 114
    Plan of Operation 115

Critical Fireground Size-Up Factors 115
    Building Characteristics 115
    Life Hazard 124
    Resources 128
    Arrival Condition Indicators 131

Summary 136

Review Questions 137

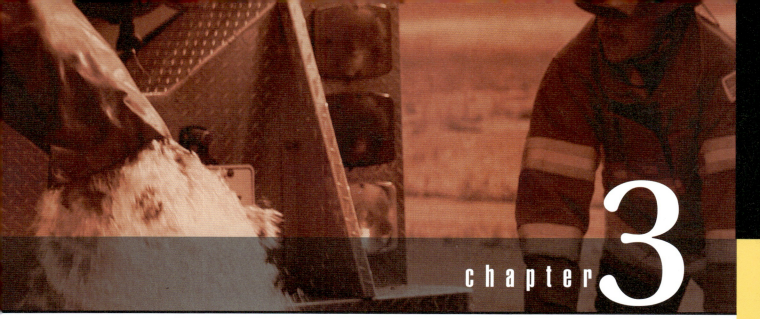

Chapter 3

Key Terms

Preincident Plan 94	Required Fire Flow 100
Preincident Survey 94	Situational Awareness 113
Q-Deck 118	Size-Up 93

Size-Up: Evaluation and Assessment

Learning Objectives

After reading this chapter, students will be able to:

1. Discuss the application of size-up theory.
2. Identify incident size-up considerations.
3. Identify critical fireground size-up factors.

Chapter 3
Size-Up: Evaluation and Assessment

In the previous chapter, you learned that different types of building construction and materials can contribute to or prevent the development and spread of fires in structures. At the same time, a building's occupancy classification also has a bearing on how fires ignite, develop, and spread. Contributing to the development and spread of fire is the combustibility of the contents in the structure. Model building and fire codes attempt to control all of these factors through both engineering and enforcement.

Model building and fire codes also require the installation of local alarm systems, monitored alarm systems, and/or fire suppression systems depending on the type of occupancy. As you know, early detection and control is critical to successfully extinguishing a fire before it does extensive damage or threatens the lives of occupants.

However, until the fire is detected and reported to the fire department, you have no control over the fire. Your job is to arrive safely on the scene, establish Command, assess the situation, develop an Incident Action Plan (IAP), allocate your resources, and control and extinguish the fire. To do this effectively and efficiently, you must collect information on the incident, evaluate it, and make the correct decision using a process known as **size-up**.

To perform an effective and thorough size-up, you should consider information that is gathered before the incident occurs, during dispatch, upon arrival, and throughout the incident. This chapter provides you with a process for gathering information. It also discusses important information that you must consider when you perform a size-up. Size-up is used at all types of incidents and not just at structure fires. The more practice you have in sizing up incidents, the better you will become at reading the visual indicators and making informed decisions.

> **Size-Up** — Ongoing mental evaluation process performed by all firefighters on the scene. Size-up is the evaluation of visual clues and collection of information in which decisions are based. It is an ongoing process as initial decisions are continually checked with new information and fire indicators to confirm initial perception and changes to conditions as the fire operations progress. Size-up results in an action plan that may be adjusted as the situation changes. It includes such factors as time, location, nature of occupancy, life hazard, exposures, property involved, nature and extent of fire, weather, and fire fighting resources.

Size-Up Application

As mentioned earlier, the size-up process actually begins before an incident is reported and continues throughout the incident. This section discusses the application of size-up theory to three specific time periods: preincident, on arrival, and during the incident.

Preincident Survey — The act of collecting information on a site prior to the occurrence of an incident.

Preincident Plan — Plan developed from the information gathered during the preincident survey and used during emergency operations and training.

Preincident

The preincident portion of the size-up process consists of the knowledge you have gained from training, study, and experience plus the site-specific information that you gain from **preincident surveys** and **preincident plans** (**Figure 3.1**). Your knowledge includes fire behavior, resource capabilities, building construction, and hazardous conditions as well as your local Incident Command System (ICS), standard operating procedures, strategy/tactics, and fireground procedures.

The preincident survey provides you with information about:

- Site access
- Structure
  — Construction
  — Occupancy type
  — Fire protection system(s)
  — Content fuel load
  — Unique characteristics or hazards
  — Access
- Available water supply
- Resource needs

Figure 3.1 Basic to all decision-making is the information gathered during preincident surveys.

Unlike fire and life-safety inspections, preincident surveys are not generally mandated by local ordinances. Preincident surveys are recommended in the following standards:

- NFPA® 1500, *Standard on Fire Department Occupational Safety and Health Program*
- NFPA® 1710, *Standard for the Organization and Deployment of Fire Suppression Operations, Emergency Medical Operations, and Special Operations to the Public by Career Fire Departments*
- NFPA® 1720, *Standard for the Organization and Deployment of Fire Suppression Operations, Emergency Medical Operations and Special Operations to the Public by Volunteer Fire Departments*

Preincident surveys are required for municipalities that adhere to the Insurance Services Office Limited (ISO) insurance rating system requirements. They may also be required by your fire department's standard operating procedures/guidelines (SOP/SOG) for specific types of occupancies.

If time and resources allow, it is good practice to perform preincident surveys on all types of structures with the exception of single-family dwellings. Residential structures may have common floor plans and characteristics that can be studied. Single-family dwellings can be informally surveyed when doing in-service home inspections and/or EMS responses. Crews should familiarize themselves with home styles, designs, and layouts common to their community or response area. Every department should have SOP/SOGs for strategies and tactics for single-family dwellings.

Company-level personnel conduct preincident surveys for the following reasons:

- Become familiar with a structure or facility, its physical layout and design, any built-in fire protection systems, and any hazardous materials, processes, or conditions that may exist. Example: a warehouse that includes high-rack storage of paper rolls which creates a high fuel load and a potential collapse hazard.

- Visualize and discuss how an emergency is likely to occur, how a fire may behave in the structure, and how existing strategies and tactics might be applied. Example: if a fire started in a building at a particular location, where to ventilate, and how to enter and attack the fire.

- Identify the number and type of resources needed to handle an incident at a particular location. Example: a quantity of hazardous materials in a pool supply store that would require specialized resources.

- Identify critical conditions that were not noted or have changed since the previous preincident surveys or fire and life safety inspections. Example: undocumented (or unpermitted) alterations to the interior layout of the facility that create a life safety hazard to occupants or firefighters.

- Identify unusual or difficult access. Example: long, narrow, or steep roadways, overhangs, weight-limited bridges/roadways, guard dogs, fortified entry doors, or other barriers to access **(Figure 3.2)**.

- Identify water supply issues. Example: lack of water distribution systems, difficult access to static water supplies, or private water supply systems.

- Build and foster good relationships with owner/occupants and residents. Example: provide information on fire prevention or contact information for important governmental services.

NFPA® 1620, *Standard for Pre-Incident Planning* provides a guideline for performing preincident surveys and developing a preincident plan. It includes a general list of conditions to look for during the survey, specific items based on the occupancy type, and samples of preincident data collection forms. A sample form is also included in Appendix A of this manual.

Figure 3.2 Site access may be restricted by barriers such as old bridges that were not designed to carry the weight of modern fire apparatus.

Preparation is essential for a thorough and efficient preincident survey. You should remember that you are not only taking your time and the time of your personnel to perform the survey but also the owner/occupant's time. Obtain a copy of the facility plot plan from your fire prevention division, the building code department, or owner/occupants. Consult the report for the last fire prevention or code enforcement inspection and preincident survey as a basis for identifying any changes or discrepancies you might encounter during the survey **(Figure 3.3, p. 96)** Finally, you must contact the owner/occupants to explain the reason for the visit and establish a mutually acceptable time.

You should follow your local SOP/SOG for determining which occupancies to survey and how often. The criteria for which structures to inspect is usually set by the potential fire and life safety hazard of the site. Referred to as target hazards, these sites contain life safety concerns for firefighters as well as occupants, hazardous processes, or storage which may have a high frequency or severity of fires, prominent structures, and/or high content or structure value. A few examples are:

Figure 3.3 Preincident surveys begin with a review of information contained in the department's facility site file.

- Life Safety Concerns
    — Schools, colleges, universities
    — Auditoriums, theaters, restaurants
    — Places of worship
    — Hospitals, nursing homes, day care centers
    — Multifamily dwellings, hotels, motels, dormitories
    — Institutions, jails, prisons
- Hazardous Processes or Storage
    — Spray paint operations
    — Metal plating
    — Chemical plants
    — Automobile fueling stations
    — Compressed gas storage
    — Paint storage warehouses
- High Contents/Structure Value
    — Mercantile occupancies
    — High-rise structures
    — Office buildings
    — Warehouses

Generally, surveys are conducted and updated annually or biannually when there has been a change of occupancy type and renovations or alterations have been made to the structure. Surveys should also be made at construction sites as a means of learning about how the building is being built. Construction sites are very vulnerable to fires and your survey is a way of preparing for that possibility.

Your department may already have a survey checklist of information you need to gather or conditions you need to be aware of. The information you gather will be used to develop a preincident or operational plan. During the survey, you should concentrate on gathering the following information:

- Life safety concerns for firefighters and occupants
- Building construction type to determine resistance to fire spread
- Building services, including utility shutoff, elevators, and HVAC
- Building access and egress
- Building age
- Building area and height
- Construction material to determine resistance to fire spread
- Contents to estimate fuel load
- Building use to determine life safety and fuel load
- Exposures
- Collapse zone
- Location and capacity of available water supply
- Location of fire control and protection system control valves and connections **(Figure 3.4)**
- Presence of existing fire control and protection systems
- Hazardous materials or processes, including flammable/combustible liquids and gases
- Location of Safety Data sheets (SDS) (also known as a Materials Safety Data Sheet [MSDS])
- High-voltage equipment
- Unprotected openings
- Overhead power lines or other obstructions
- Occupancy load at all hours
- Names and telephone numbers of contact or responsible persons for owner/occupant
- Estimated quantity of water required to extinguish a fire in the structure or a portion of it (fire flow)
- Emergency evacuation plan

Figure 3.4 The preincident survey should include all fire protection systems including suppression systems, shutoff valves, and risers.

Preincident Survey Process

The preincident survey process is usually based on an established procedure defined by your department and will vary between organizations and jurisdictions. The process outlined here is based on generally accepted practices used by fire departments nationwide.

As already explained, the first step is to contact the owner/occupant, explain the reason for the survey, and determine a convenient time for the survey. When making a preincident survey, you should meet the owner/occupant or a designated representative before beginning the survey. Explain that the preincident survey is not a code enforcement inspection; however, any serious fire or life-safety hazards found will need to be corrected. When serious hazards are found, the best approach is to attempt to obtain an on-the-spot correction, and follow up with the appropriate method such as a memo to the fire prevention and inspection division. If compliance cannot be obtained voluntarily, legal means should be used to require compliance. You should explain that the preincident survey is also a benefit to the owner/occupant because it may prevent the temporary closing of a business resulting from an emergency at the site, minimize fire suppression damage, and safeguard lives. For serious life-safety hazards, the fire prevention personnel may be requested for immediate corrective action of such structures.

Preincident surveys should be made in a systematic and logical approach. Generally, the survey is divided between the exterior and interior of the structure. At a large facility such as a chemical plant, the survey may start with the exterior of the main plant building, proceed to the interior of the building, and then extend to other surrounding structures on the site (**Figures 3.5a & b**).

As previously stated, the owner/occupant or a designated representative must accompany you during the survey. This procedure ensures that you will be able to gain access to areas that are normally secured, have someone to answer questions, and prevent you from accruing any liability. Equally important, this allows fire service personnel an opportunity to educate owners/occupants on best practices in fire prevention/life safety awareness. You may encounter areas that are locked or contain processes that require certain precautions, such as wearing chemical suits for entry into the area. The person with you can either explain the situation or arrange for access.

During the survey, you should look at ventilation systems, fire protection systems, water supplies, structural conditions, content fuel loading, best access points and estimated lengths for hoselines, and methods for property conservation. The location of utility shutoffs, sprinkler control valves, and passive ventilation controls should be determined as well as the means for operating them. The locations should be noted on the site plan and described on the survey form.

Survey information may be gathered in a variety of methods, including any combination of checklists, written essay-style commentaries, sketches, site plans, photographs, and videos. You should be thorough in collecting the information. The bulk of the information should be kept in a facility file and shared with other units that will respond to the site and/or with your fire prevention and code enforcement division.

While the facility file should be detailed, the information used for the preincident plan should be limited to the essential data. Preincident plans, either in hard copy form or on a mobile data terminal in the apparatus, typically should not be more than two pages long including the site plan. Some large complexes may require more than two pages. (**NOTE:** Responders must be able to read the plan quickly while en route to the incident.)

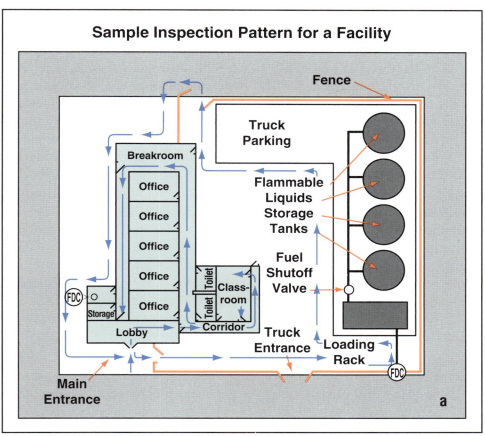
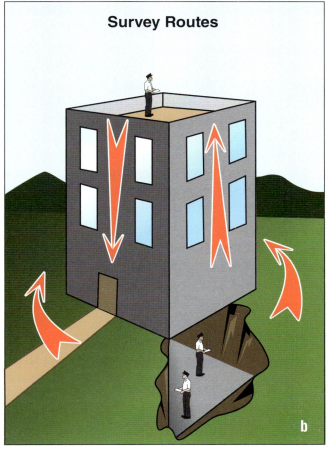

Figures 3.5a&b Illustration (a): example of a survey pattern for a large facility; Illustration (b): example of a survey pattern of a multistory building.

The preincident or operational plan should contain enough information for the Initial Incident Commander (IC) and all subsequent officers to have an accurate picture of conditions in the structure at the time of the survey. The plan should be easy to read, accessible, timely, and accurate. These plans provide facts based on the survey that will help personnel make decisions during an incident at any of the surveyed locations. Regardless of whether preincident plans have been developed for a particular site, the size-up process still begins before an alarm is sounded and continues during the response. Preincident plans can also be used for company training evolutions and training scenarios.

Required Fire Flow Calculations

Part of the preincident survey involves calculating the required fire flow for the structure. The term **required fire flow** is used to describe the estimated uninterrupted quantity of water expressed in gallons per minute (gpm) (liters per minute [L/min]) that is needed to extinguish a well-established fire. It is important that you be able to calculate the quantity of water that will be needed to extinguish a compartment or structure fire. Knowing the amount of water needed will help determine if you have the resources available to apply the water, including staffing and pump capacity **(Figure 3.6)**.

> **Required Fire Flow** — The estimated uninterrupted quantity of water expressed in gallons per minute (gpm) (liters per minute [L/min]) that is needed to extinguish a fire.

Figure 3.6 Knowing the fire flow required for a structure will help to determine if there will be resources available including water supply, personnel, and time, as in the case of a drafting operation.

You should calculate the fire flow for each type of structure within your response area during the preincident survey of target hazards or other structures. Some structures, such as single-family dwellings of similar floor space (area), can be based on one general fire flow requirement because they are similar in construction and fuel load.

Numerous formulas exist for calculating required fire flow, and some are more intricate than others. Those formulas based on detailed mathematical formulas, such as the Insurance Services Office (ISO), International Code Council (ICC), and Factory Mutual (FM) formulas, are intended for use by insurance and building officials. In addition, NFPA® 1142, *Standard on Water Supplies for Suburban and Rural Fire Fighting*, contains a fire flow formula that includes occupancy and construction types as well as considerations for exposures.

However, simpler formulas are available that will provide an adequate estimate. It is important to know the formulas used to calculate and determine the needed fire flow for any given structure. By taking the time to calculate the required fire flow during the preincident survey and including that information in the preincident plan, you will save critical time during an emergency incident.

In the 1980s, the National Fire Academy (NFA), developed a formula for estimating fire flow based on the percent of involvement of the structure or compartment. This formula was intended for both preincident planning and for on-scene calculations at the incident:

Needed Fire Flow (NFF) = (Length x Width)/3 x % involvement

100 Chapter 3 • Size-Up: Evaluation and Assessment

This formula is intended for use with an offensive interior fire attack in a compartment or structure that is no more than 50% involved. The formula is less accurate, and therefore less effective, at involvement greater than 50% or when flows are determined to be greater than 1,000 gpm (3 785 L/min). Pre-incident estimates using this formula are generally calculated for 25%, 50%, 75%, and 100% involvement.

Calculating Fire Flow

Although there are numerous formulas to calculate fire flows, the initial Incident Commander usually will not have time to use them during a structure fire. For situations where you do not have a fire flow estimate or where the fire has exceeded the estimate, apply the simple rule of thumb. You can also apply the statement, "Big Fire, Big Water." That is, the larger the fire, the more water, stated in GPMs (L/min), that will be required to extinguish it. In any case, it is better to be prepared with sufficient supply hoselines and attack lines in place than to not have enough capacity available **(Figure 3.7)**.

Figure 3.7 A major defensive strategy may require multiple master streams, apparatus, hoselines, and personnel. *Courtesy of Tom Aurnhammer.*

Before an Alarm

As you travel to work, you probably begin a general size-up for the day. You may observe potential barriers, obstructions, or conditions that will slow or prevent a safe, rapid response. These may include road maintenance or construction areas as well as designated detours. You should review the day's weather forecast and ask yourself how and to what extent the wind and temperature might affect any emergency incident. Finally, you should consider answers to the following questions:

- Will response time be slowed because rain, snow, or ice has made the roads slick and dangerous?

- Will ventilation crews be at additional risk because of wet or icy roofs or high winds?

- Will extreme weather adversely affect trapped or injured victims?
- Are there any extremes of temperature that could make it more difficult and perhaps dangerous for personnel?
- Will wind velocity and direction affect vertical or horizontal ventilation?
- Are there any training activities scheduled that may put responders out of position or cause them unusual stress or fatigue prior to an emergency occurring?
- What are the locations of school zones and bus stops where additional caution is required during loading and unloading?
- Will detours be necessary because of construction or other factors — parades, demonstrations, weight limitations on bridges, railroad crossings, etc.?

Emergency responders performing strenuous work while wearing protective clothing in hot weather can be especially vulnerable to heat-related illnesses such as heat exhaustion or even heatstroke. You should be aware of the possible effects of weather and other factors on the abilities of your personnel to perform effectively and safely.

While Responding

The size-up process continues when the alarm is sounded or your pager is activated. Depending on the dispatch system that your department uses, listen carefully to all information provided through the radio broadcast **(Figure 3.8)**. If information is provided in a printed message or on a computer terminal in the apparatus, read it carefully. This information contains a description of the current conditions as reported by witnesses at the scene. You must remember that not all of the information given when dispatched may be factual and could be highly inaccurate and incomplete.

Figure 3.8 You must listen to the dispatch report for all essential information about the incident.

Prior to the arrival of fire personnel, the only thing you can verify immediately is the time of day, weather conditions, and capabilities of your department's response. You must be prepared for any type of situation in which you may be faced. For instance, is the caller at, near, or across from the address given? Is the report of a building filled with smoke actually the odor of a smoldering fluorescent light ballast? Even law enforcement officers who are not trained as firefighters have been known to announce a building to be fully involved in fire that later proved to be only a small kitchen fire. Conversely, reported vehicle fires have been found to be well-involved vehicles parked in garages under apartment buildings.

Pay particular attention to nonstandard alarm assignments. For example, if units are being called that are not normally on your first assignment, or if some units are missing which may indicate that some resources are unavailable. Local SOP/SOGs will generally establish the minimum information that is given during a dispatch broadcast, including:

- Time — Exact time of dispatch, not the time the fire started
- Situation — Type of emergency reported to the dispatch center

- Location — Address of the emergency situation
- Resources dispatched — Units responding

Time is an important factor in the size-up process. Although you do not initially know the time the fire started and how long it has been burning, being able to calculate your response time and the visual clues of the fire upon arrival will help determine the stage of the fire.

You should also consider the time of day when the alarm is received and how it will affect the incident by considering the answers to the following questions:

- Is it during normal business hours when commercial properties are occupied?
- Is it during evenings or weekends when residential properties are more likely to be occupied?
- Is it the middle of the night when residential occupants may be sleeping?
- Is it during school hours on a weekday?
- How are the month, day of the week, and time of day likely to affect traffic congestion along the response route?

You should continue to evaluate variables as you respond to the reported incident. Consider both the weather and the capabilities of your resources by asking the following questions:

- Weather:
  - How will the weather affect reflex and response times?
  - How will the weather affect apparatus safety during response and set-up?
  - What affect will the weather have on visibility?
  - What affect will the weather have on fire conditions?
  - How will the weather affect firefighter safety?
- Resource capabilities:
  - What is my physical condition and that of my crew?
  - Do I have all of my regularly assigned crew?
  - How long before additional resources arrive at the scene?
  - Have the correct resources been dispatched to meet the current hazard?
  - Are my resources trained to meet the actual hazard?
  - Where is my water supply and is it sufficient to control the situation?
  - Are my personnel equipped and trained to operate in an immediately dangerous to life or health (IDLH) atmosphere?

Additionally, consider any information that you have available. For example, the following information should be considered at a fire incident:

- Review the preincident plan for the building (if available) to more thoroughly prepare for what responders may encounter.
- Observe the color, thickness, volume, velocity, density, pressure, and movement of any smoke produced by the fire.

This information, combined with knowledge of fire behavior and the building where the fire is burning, can help you better assess resource needs. You should also ask for and evaluate any additional information provided by the dispatch center over the radio during the response. Based on knowledge of the response area, the units assigned, and the radio communications, you should be able to determine which unit will arrive first and establish the ICS.

Facts, Perceptions, and Projections

When considering the foundation for your decision-making process, think about three components: facts, perceptions, and projections. If you learn what these are and how to recognize them, you will be able to make sound decisions when you arrive at the incident scene.

Facts

Facts are the things that you know to be true. Facts are based on your preincident survey of the site (if one has been performed and recorded), knowledge of building construction and occupancy, the current time of day and the day of week, and on-scene observations. Knowledge of basic fire development as well as an assessment of your resources will also play an important role in determining your course of action.

Site knowledge — Your preincident survey should give you firsthand knowledge of:

- Potential hazards faced by occupants and firefighters
- Where occupants may be located
- Evacuation meeting site for occupant accountability purposes
- Expected number of occupants in the structure at any given time
- Paths of egress and potential barriers to egress
- Access and potential barriers to access
- Areas that can be used to shelter-in-place
- Type of construction
- Type of occupancy and use of the structure
- Physical layout of the structure
- Water supply source(s)

You should also consider that a preincident plan may not exist and that you may not have made a survey of the site. In these cases, your facts may be reduced to a minimum and based on the information provided in the alarm dispatch and what you see upon arrival.

Construction and occupancy — Knowing the type of construction can help you project potential fire growth, potential directions of fire travel, and ways to use the structure to protect occupants. Knowledge of the type of occupancy or use of the structure can tell you the potential number and location of occupants. It can also alert you to the ability of the occupants to self-evacuate or if they will require assistance. Construction and occupancy can also give the first-arriving officer an idea of the dangers and obstacles the unit may face in trying to attack the fire and perform rescue such as steel bars in windows and doors creating problems for occupant egress and firefighter access. Occupancy information may indicate possible hazards while searching. Construction type and occupancy dictates strategy and tactics (e.g. a concrete roof delays or eliminates the possibility of vertical ventilation).

Time of day/day of week — Although some types of occupancies are inhabited 24 hours a day, 7 days a week, most are not. The presence of security personnel or a cleaning crew is always possible and must be considered. The preincident survey may include this type of information.

Fire behavior — The ability to estimate how fires will grow, spread, and generate toxic atmospheres within a given structure or occupancy will help determine how much danger the occupants are in and the level of risk to which your crew will be exposed. In some incidents, such as the MGM Grand fire discussed in Chapter 2, fatalities caused by smoke inhalation occurred on floors far above the actual seat of the fire. Fire spread determination/calculations occur in controlled settings. The only real assumption you can reasonably make is that the fire will spread regardless of the construction if water is not applied correctly, sufficiently, and swiftly.

Resources — You must have appropriate resources to attempt any action. Resources include sufficiently trained personnel with appropriate equipment and water supply.

Perceptions

Your perceptions are based on observations influenced by your knowledge, biases, beliefs, and past experiences. What you perceive may not be based on facts at all. That is, you may "see" what you want to see and not what is actually present. The presence of lights on or vehicles in the parking lot of a commercial building at 2:00 a.m., children's toys near the front door of a residence, or evidence of forced entry to a vacant building may be indications that the building is occupied. These indications can be used to justify implementing a rescue tactic. Your perceptions may also be influenced by your emotions. In a small community, you may personally know the people who live or work in the structure. Hysterical neighbors or witnesses who have no actual knowledge of who is in the structure may influence your decisions. Perceptions are important, but they should not take precedence over the facts as you know them.

Projections

Your ability to predict or project what might happen next is based on your knowledge of fire behavior, building construction, and fire fighting activities.

- ***Fire behavior*** — Rapid fire development, fire growth, and fire spread may threaten emergency responders, occupants in other portions of the structure, and adjacent structures. Having resources available to address changes in fire growth is important. It is desirable to have resources staged and not committed to a task.

- ***Building construction*** — Knowledge of building construction can help you predict how long a fire can be contained within a compartment and how rapidly it can spread through concealed spaces. This knowledge also helps to estimate when the structure, or a portion of it, has the potential to collapse. Knowing the location of access and egress doors and stairways will assist in predicting areas to expect victims, searching for and removing victims, protecting yourself and your crew during the search operation, advancing hoselines, and performing ventilation.

- **Fire fighting activities** — Improper application of fireground activities can increase fire growth. Improperly performed tasks such as nozzle pattern selection, inadequate fire flow, attacking from the wrong direction, and improperly coordinated ventilation will cause the fire to spread, endanger lives, and may cause a flashover or backdraft. Coordinated ventilation and fire attack must be monitored. The knowledge of local fireground procedures and activities can help you project the potential need for a firefighter rescue.

Firefighter fatalities at structural fire incidents tend to be the result of personnel becoming disoriented and running out of air or being involved in a structural collapse. Creating and maintaining multiple means of egress, use of hoselines or tag lines, proper ventilation, and proper air management are critical to firefighter safety. Knowing where crews are at all times through an effective accountability program, having an established rapid intervention crew/team (RIC/RIT) on site, keeping in constant communication with all crews by requesting continuous feedback reports, and being aware of how the fire is developing are methods of preparing for a potential rescue.

The Federal Occupational Safety and Health Administration (OSHA), some state regulations, and NFPA® standards mandate the presence of an RIC/RIT outside of a burning structure for the potential rescue of firefighters. While the regulation only requires two trained personnel as RIC/RIT, it is highly recommended that the crew be at least four members and include an officer. In a true RIC/RIT deployment, it will take several crews to facilitate a rescue. The IC must use the same care in planning a rescue effort as planning for fire control. Sufficient and appropriate resources must be assigned to safely carry out the rescue assignment. While the NFPA® standards and federal/state/provincial regulations allow the RIC/RIT personnel to have concurrent outside assignments, those assignments may not be such that leaving them to assist the firefighters inside would increase the jeopardy of any firefighters on the scene.

On Arrival

The most intense part of the size-up process occurs when you arrive at the emergency incident scene. You may be greeted with a scene of utter chaos or one with no visible clues that an emergency exists. In addition to the emergency situation and those individuals directly involved in it, numerous spectators may be gathered at the scene, making it difficult to distinguish them from occupants or victims. These bystanders may be hysterical or irrational and screaming for responders to do something. Some may be attempting to extinguish the fire, assist victims, or perform a rescue — actions that may place them in direct danger.

In the midst of this scene, you must assess and communicate critical information to other responding units and the dispatch center. Your scene assessment begins before the apparatus comes to a stop and you step from the apparatus. The incident may be categorized by the severity, extent, and dynamics of the incident. Two general situations are encountered upon arrival. There may be nothing evident which will prompt an investigation, or there may be smoke or fire showing, prompting an attack **(Figures 3.9a & b)**.

Figures 3.9a&b Upon arrival, the initial Incident Commander may select (a) investigation or (b) attack depending on the immediate conditions.

These two situations are the core of the arrival or condition report (conditions on arrival or size-up report in some jurisdictions) that you broadcast over the radio. Many fire and emergency services organizations mandate this report in their operations SOP/SOG. See Appendix B for a sample operations SOP/SOG. This initial report is an essential communication that will inform the other responders and establish a foundation upon which to build the incident.

The arrival report contains:

- Unit arriving on scene
- Correct address of the incident if different from the dispatch
- Description of the structure including:
  — Number of floors
  — Type of construction
  — Type of occupancy
- Description of conditions found at the scene
- Special considerations
- Operational strategy
- Intended initial actions
- Water supply need or availability
- Incident Command decisions
- Requests for any additional resources

The report provides other responding units with an idea of what they will encounter — you are describing the fire scene for crews who have not arrived yet. It should clearly portray what is to be expected by giving a description of the scene including building type and size, extent and location of fire, life safety issues, occupancy type, and water supply needs or availability. The report can also include the type and location of any special hazards such as barriers that could impede access, the location and condition of victims in need of assistance, downed power lines, and other important observations. The arrival report also includes a Command statement, the location of the Command Post, a description of initial action, and the Command option (investigation, fast-attack, command) that the officer is planning to use. If the operation is going to be managed over a specific radio frequency (some jurisdictions assign a tactical channel on dispatch), then this information is also provided to responding units and they are directed to move to that frequency.

Depending upon the organization's SOP/SOG, if responders from the first-arriving engine do not deploy a supply hoseline from the nearest accessible hydrant (or otherwise ensure a water supply), responders from the next arriving engine may be assigned this operation **(Figure 3.10)**. As with any assignment, you as the initial IC must be specific about assigning water supply as in ordering, "Engine Two secure a water supply on your arrival." Again depending upon policy, this procedure may be automatic or it may have to be assigned by the IC. In the latter case, providing the water supply would be part of the IAP and one more decision that the first-arriving officer must assign.

Following the arrival report, perform a visual size-up by looking at the scene from all sides. If you cannot physically complete a full 360° survey or walk around the structure, you must use other means to learn conditions on all sides of the incident **(Figure 3.11)**.

While checking all sides of the structure, information you are gathering includes:

- Any indication of occupants or the need for rescue
- Any indicators of fire location
- Lowest floor of fire involvement
- Smoke conditions
- Forcible entry requirements
- Special hazards: propane, chemicals, electrical lines, elevation changes
- Building construction

Figure 3.10 Department policy and procedures determine whether the first pumper will provide its own supply line or if a later arriving unit will set up to pump from a hydrant.

Figure 3.11 When the situation allows, the IC should perform a 360° survey.

When the structure is small, such as a single-family dwelling, you may be able to make a 360° survey. That is, walk all the way around the structure to view it from four sides. In the case of large structures, strip malls, row houses, structures with limited access, or big box retail stores, personally performing a 360° survey may not be possible. In these instances, it will be necessary to have other members of your crew or other arriving units provide information from the other sides. In commercial buildings, the roof is often included in this request and aerial apparatus or ground ladders are used to view the roof area. Other incoming units may be assigned to various sides of the structure and supply a report back to the Incident Commander **(Figure 3.12)**.

You must focus on the situation and answer the question, *Can the resources at the scene and en route handle this situation?* If the answer is *no*, or even *maybe*, then additional resources must be requested *immediately*. Addressing this critical question is why the initial size-up sets the tone for the balance of the incident.

If there are not enough resources initially available and the arrival of additional ones may be delayed, personnel are likely to start the incident *behind the curve.* This type of situation will take immense work and organization to overcome. Therefore, it is critically important for the initial size-up to be done with attention to detail and focus on the desired outcome and the activities and resources needed to accomplish that outcome. Considerations must be given to the level of acceptable personnel risk (that is, risk/benefit analysis) and which operational strategy to initiate. Properly interpreting condition indicators can help you make the correct decisions.

Once the 360° survey is performed, determine the appropriate course of action, develop an Incident Action Plan (IAP), and assign duties to your crew and the remainder of your resources. Keep in mind that additional units may not

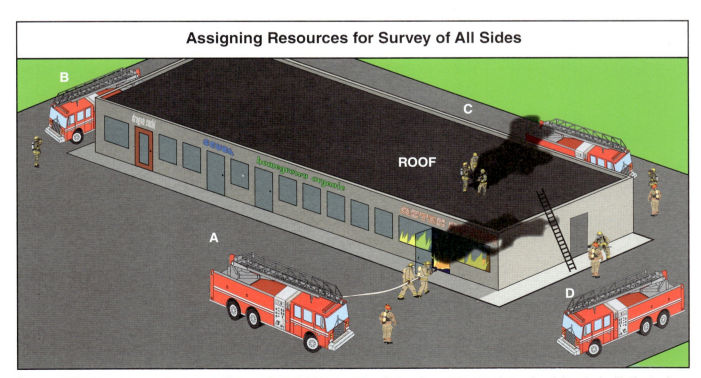

Figure 3.12 If it is not possible for the IC to perform a 360° survey, other units or personnel may be delegated with the task of reporting on conditions from other sides of the building.

arrive for some time, and it may be necessary for you to prioritize the activities your plan requires. You must also keep in mind that while you are setting up operations, the fire is increasing in size and extension. A risk assessment is necessary to ensure that you are making the right choice and doing it safely. (**NOTE:** While your crew may assist with information collection and offer suggestions, the Incident Commander is responsible to make all decisions.)

Before you commit any resources to an interior fire attack, you must have the resources to adhere to the 2-in, 2-out rule. For the first two firefighters to enter the IDLH, you must have two on the outside prepared to respond to a firefighter rescue. The two personnel on the exterior must be fully trained and prepared to enter the structure in full personal protective equipment (PPE) including SCBA. These individuals may be the pump operator, yourself, or another firefighter. If you cannot implement the 2-in, 2-out rule, you must only send personnel into the structure if there is a known rescue, an incipient stage fire, and/or the fire can be extinguished with a portable fire extinguisher.

NOTE: The presence of four firefighters at a structure fire does not automatically ensure that the 2-in, 2-out rule has been met. If it is necessary for one or both of the outside members to leave their primary duty, they may not be considered available for rescue. In other words, they may not leave their post if it will further endanger the incident scene. A member performing a task critical to the safety of other members on the scene should not be considered available for the initial rapid intervention team if by leaving that assigned task to make a rescue, other on-scene members are placed at greater risk. Your decisions must be driven by your local SOP/SOG and by best industry practices if national, state/provincial, or local laws/ordinances do not apply.

Occupational Safety and Health Administration (OSHA) Two-In, Two-Out Rule

The following is the applicable portion of OSHA 1910.134(g)(4) used to establish the Two-In, Two-Out rule:

Procedures for interior structural firefighting. In addition to the requirements set forth under paragraph (g)(3), in interior structural fires, the employer shall ensure that:

1910.134(g)(4)(i)
At least two employees enter the IDLH atmosphere and remain in visual or voice contact with one another at all times;

1910.134(g)(4)(ii)
At least two employees are located outside the IDLH atmosphere; and

1910.134(g)(4)(iii)
All employees engaged in interior structural firefighting use SCBAs.
Note 1 to paragraph (g): One of the two individuals located outside the IDLH atmosphere may be assigned to an additional role, such as Incident Commander in charge of the emergency or safety officer, so long as this individual is able to perform assistance or rescue activities without jeopardizing the safety or health of any firefighter working at the incident.

Note 2 to paragraph (g): Nothing in this section is meant to preclude firefighters from performing emergency rescue activities before an entire team has assembled.

During the Incident

After the Incident Action Plan is implemented, emergency responders will be busy performing their assignments and making progress toward resolving the incident. This phase (between arrival and termination) can be relatively short, or it can last for a considerable length of time. When the problem is simple or small and/or you make good decisions, the incident may be resolved in a few minutes. If the problem is complex, additional resources and assistance from other officers may be needed to resolve the situation.

During this phase, the situation changes by either improving or worsening. In either case, decisions that were based on the initial size-up may or may not remain valid. Therefore, constant reassessment or ongoing size-up of incident conditions and the effect that operations have on the problem is critical. You must continue to size up the situation; validate or adjust the incident objectives, strategies, and tactics; and make changes to the IAP, as needed. By getting continuous feedback from crews, you can follow the progress of your IAP and continue to size up the scene. An acronym for this part of the size-up is CARA, which stands for:

- *Conditions* — An assessment of the interior conditions and the visible indicators on each side and roof of the structure
- *Actions* — Specific activities being performed by each crew inside and outside the structure that are engaged in fire fighting operations
- *Resources* — Includes additional personnel, apparatus, equipment, ventilation, a second hoseline, or an exchange of personnel
- *Air* — Refers to the breathing air supply available for interior fire attack as well as salvage and overhaul operations

Variations on CARA

Many regions, states/provinces, and local departments use variations on the CARA acronym. What you use should be based on your own SOP/SOG. Two other variations include:

CAN	CAAN
Conditions	Conditions
Actions	Actions
Needs	Air
	Needs

Incident Size-Up Considerations

An effective company officer is someone who not only can determine whether additional resources are needed but also determine the number and type of resources that will be needed. The reflex time is a factor in this determination (**Figure 3.13, p. 112**). Once on the scene, you must be proficient at assessing how the incident is developing, how rapidly it is expanding, and where it will be in both intensity and location prior to additional resources becoming operational at the scene. You must understand that implementation of your IAP

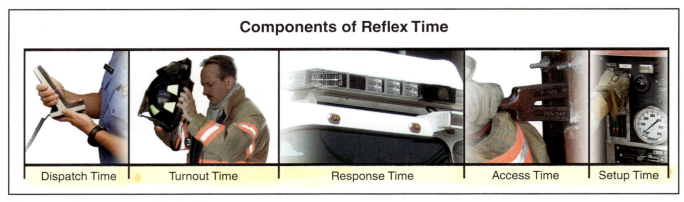

Figure 3.13 Components of Reflex Time.

will take time. The resources required to implement the plan need to arrive on scene. The units already on scene will take time to put on all their PPE, deploy attack and supply hoselines, and set ground ladders. This delay time needs to be considered because the fire will continue to grow until crews are operating and water is applied to the fire.

A number of size-up processes or models are available for you to use. One of the accepted models was developed by Chief Lloyd Layman in the 1940s. Decades after Chief Layman wrote his seminal work *Fire Fighting Tactics*, its principles are as valid as ever, and his traditional model is the one most commonly used today. In his book, Layman described the following considerations needed for analyzing any emergency situation:

- *Facts* — Things that are true
- *Probabilities* — Things that are likely to happen
- *Own situation* — Your knowledge about the situation and an estimate of what can be accomplished with the available resources
- *Decision* — Initial use of resources followed by supplemental resource needs: Clear and precise outline of the actions to be taken and the objectives to be achieved
- *Plan of operation* — Information compiled into an Incident Action Plan (IAP)

Facts

The facts of the situation are things that you know to be true and are actually observing. Some of this information may be provided by the dispatch center based on the report of the emergency. Remember that, as mentioned earlier in this chapter, not all the information given when dispatched may be factual. All of this information can and should be factored into your thought process regarding the emergency. Some of the information includes the following:

- Time (month, day, hour)
- Location (address, business name, landmarks)
- Nature of the emergency (fire, hazardous materials release, structural collapse, motor vehicle accident, medical emergency, etc.)
- Life hazard (occupants and responders)
- Exposures (adjacent uninvolved property)

- Weather (wind, temperature extremes, humidity, etc.)
- Estimated or confirmed number of trapped or injured victims
- Resources being dispatched including personnel

Not everything you see can be called a fact. Sometimes you can be misled to accept as a fact what is really only a perception. A perception is based not only on what your senses indicate but on what your training, experience, and even bias tell you that you *should* see. For instance, most firefighters associate a specific odor with a fluorescent light ballast that has overheated and melted. The association between the odor and the cause can be so strong that you may ignore other indicators of the actual hazard or condition that is causing the odor. While this association can also lead to solving the problem quicker, you should be aware that there may be another cause for the situation.

To counteract misperceptions, you must rely on additional sources for your gathering of facts. One of the best resources is the observations of your crew members. You should encourage your crew to apply the concept of **situational awareness** during emergencies. Communicate with and listen to the observations provided by them throughout the size-up. The synergy created by the shared knowledge and experience of the entire unit should provide a balanced approach to decision-making.

Situational Awareness — Being aware of your surroundings and what is going on.

Probabilities

Probabilities are things that are not known for certain, but based on the facts that are known. They are things that are *likely* to happen. To assist in making decisions, the following questions must be answered regarding the probabilities of a fire-emergency situation:

- Where is the fire located?
- In which direction is the fire likely to spread, given building layout and configuration, types of fuel, and fire behavior?
- Are exposures at risk?
- Are explosions likely and is a secondary explosion likely? Is structural collapse likely?
- Do occupants need to be evacuated?
- Are additional resources likely to be needed? If so, what types and how many?

Many of the decisions involved in the probabilities phase of a fire incident size-up can be made easier and the result more accurate when you have some knowledge of the following factors:

- Fire behavior and smoke indicators (from past experience, training, and education)
- Building construction type and material involved (from knowledge, training, and preincident planning)

This knowledge is especially important if the building involved in fire has lightweight truss construction. Many modern buildings have lightweight wood or metal truss components that have a tendency to fail early in a fire, creating a significant risk of early collapse for first-arriving crews. However, the mere presence of lightweight construction trusses does not mean a collapse is imminent. As mentioned in Chapter 2, the two types that are most commonly

Figures 3.14a&b Lightweight construction trusses may be constructed of (a) metal or (b) wood.

encountered are lightweight metal bar joist trusses and wood trusses that use glue or metal gusset plates in their construction (**Figures 3.14a & b**). If a fire does not exist in an attic or cockloft area where wood trusses are found, then they may not be a concern at all. The dangers of floor truss and wood I-beams is probably a much more serious issue when sending firefighters to floors where fire may be under them. This is why determining location of fire origin and not where it is presently visible is so important. Floor trusses also create void spaces that form avenues for fire spread. Wood and composite I-beams are a commonly involved structural component involved in floor collapse in recent years.

Own Situation

Your own situation is one set of facts that is known about the overall incident situation. The following facts are among those to consider:

- Number and types of resources responding to or already at the scene
- Additional resources that are available immediately, with some delay, and with considerable delay
- Capabilities and limitations of resources (important factors in the development of an IAP)
- Assessment of your ability to deal with the situation based on training and experience
- Sources for water supply including apparatus tank, water tender, relay pumping, water supply system, or static water supplies

Decision-Making

Using the information gathered upon arrival and from the 360° survey, you have established an action plan. You have identified the problem and selected the incident priorities based on life safety, incident stabilization, and property conservation (LIP). Your IAP contains a clear and precise outline of the actions to be taken and the objectives to be achieved.

Chief Layman identified two or more separate decisions that must be made in the ongoing size-up process — an initial decision and one or more supplemental decisions based on the three incident priorities of LIP. The initial decision may be seen as having the following three segments:

1. Whether resources at the scene and those en route are adequate for the situation
2. How to deploy the resources already at the scene in the most effective manner
3. What to do with the resources that arrive (immediate deployment or staging)

As the incident progresses and the situation changes, supplemental decisions will have to be made. For instance, the IC needs to decide whether the initial deployment of resources is still producing the desired results or if the deployments need to be changed. As the fire scene progresses and the situation changes, new objectives are formed or lower priority ones can move up the list to be accomplished. Also, on very large incidents, consideration must be given to relief personnel, additional supplies, rehab, etc.

Plan of Operation

As previously mentioned, the plan of operation need not be in writing on relatively small, routine incidents involving only an initial assignment. However, there should be a written IAP for long duration or complex incidents. A written IAP is very useful when command changes or multiple jurisdictions are involved. How the plan is implemented is the operational phase. The plan is started by assigning tactical objectives to company officers in Divisions or Groups. Select the highest priority assigned first and work down your list as more crews arrive or on-scene crews complete tasks.

Critical Fireground Size-Up Factors

What you look for in your size-up is determined by the critical factors required to make your decisions. Many departments include a list of factors in their SOP/SOG for fireground operations. These factors include:

- Building characteristics
- Life hazard
- Resources
- Miscellaneous conditions
- Arrival condition indicators

There is no specific order that you should follow as you consider these factors. Some will be based on your preincident plan and personal knowledge of the site. Each factor, however, will have an effect on the decisions you make and the actions you take.

Building Characteristics

Building characteristics are contained in your preincident plan, if one exists, and supplemented by your knowledge of building construction. The preincident plan should at least provide the following information:

- Occupancy type
- Construction type
- Floor area
- Ceiling height

- Number of stories
- Contents
- External exposures
- Special hazards
- Hours of operation
- Location of water supply
- Location of utility shutoffs
- Location of fire protection system controls

If a preincident plan is not available, your initial information will come from your first view of the exterior of the structure. This view should tell you:

- ***Access points*** — Types and location of doors, windows, exterior stairs, basement entrances, garage doors, skylights, and other openings
- ***Barriers to access or egress*** — Security bars, fences, gates, and other physical barriers to access **(Figure 3.15)**
- ***Barriers to aerial operations*** — Overhead wires, trees, or porte cocheres will prevent or obstruct aerial apparatus operations
- ***Topography*** — Slopes that may indicate grade-level access onto different floors from alternate sides of the structure
- ***Roof design*** — Type of roof, peaks, dormers, overhangs, potential concealed spaces, passive ventilation systems, skylights, chimneys to indicate location of furnace, and vent pipe to indicate kitchen and bathroom locations
- ***Construction type*** — Type of exterior wall construction which can help determine the overall construction type
- ***Number of stories*** — Floors above and below grade
- ***Exposures*** — Exterior proximity to other structures, existence of fire separation walls
- ***Area*** — Approximate floor space
- ***Height*** — Ceiling height of ground level floor
- ***Occupancy*** — Determined by building appearance or signage
- ***Fire protection systems*** — Signs indicating location of FDC, fire control room, or private water supply system

Figure 3.15 Streets or pedestrian malls may be blocked to apparatus access around certain buildings.

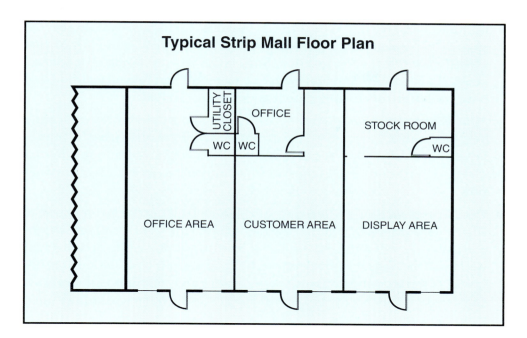

Figure 3.16 Typical floor plan for strip malls.

Knowledge of local building styles can also assist in determining the interior layout of the structure. For instance, neighborhoods with two-story homes all designed and built alike will generally have similar interior floor plans. Commercial occupancies in a strip mall will generally have a showroom or display area in the front, storage at the rear, and small areas used for toilets, basement access, and offices near the back **(Figure 3.16)**. Strip malls usually have glass store fronts and a steel door without exterior door handles in the rear. In areas of high crime rates, security grilles, gates, roll-down doors, or fortified entrances will cover windows and doors on residences and front and rear doors of commercial buildings. These obstructions will not only slow access, they will create egress hazards for occupants and firefighters who are inside the structure. In some communities, thick iron plates have been installed on the exposed flat roofs of commercial buildings to prevent people from cutting holes for illegal entry. These plates will prevent ventilation and pose a serious collapse hazard.

By knowing building construction, you will be able to make decisions based on the type of construction such as:

- *Type I construction* — Provides 3- to 4-hour fire resistance, limiting fire spread **(Figure 3.17)**. Type I structures may have fewer exits to the outside. Older Type I buildings may be more compartmentalized.

- *Type II construction* — Provides 0- to 2-hour fire resistance creating very short interior operating times. There may be a lot of exposed unprotected steel which can result in early collapse. Ceiling/roof bar joists can be hidden by drop ceilings.

- *Type III construction* — Structural members can fail between 10 (exposed wood or metal bar joists) and 20 (ordinary wood construction) minutes. There is some degree of compartmentation consisting of

Figure 3.17 Concrete panels, either poured in place or at a plant, can be tilted up to form the walls of Type I structures.

multiple load-bearing walls. There is potential for hidden fires due to void spaces. These spaces tend to show smoke from many areas with no real identification of the fire location. Strip malls with common attic spaces, heavy floor loading, and fuel loads in basements can be well fortified and have backdraft/or advanced fire conditions on arrival.

- ***Type IV construction*** — The use of large dimensional timbers will create long burn times. Large open spaces may exist in structures used for manufacturing, storage, etc. These may have been converted into residential dwellings presenting the same void space concerns as Type III construction.

- ***Type V construction*** — The time available for an interior fire attack or rescue will be determined by whether the fire involves the structure. Lightweight construction will reduce the interior operating time while dimensional wood will increase the interior time. Balloon-frame construction aids fire spread to the attic and into void spaces in floor and ceiling assemblies. This construction can usually be identified by long and narrow windows lined up in the same stud bays.

Knowledge of roof types and styles offer the following information:

- ***General*** — Roof type can either inhibit or accelerate fire spread making a roof report essential. The roof report should indicate whether the roof can be safely and effectively vented. If you are concerned about potential roof collapse, then it is probably not safe to be under it. Fire extension will be expected and determined by the smoke coming from under the roof area, peak vents, or eaves. Smoking shingles, melting/bubbling tar, evaporating or melted snow all indicate heat underneath, although attic insulation can hide this. Fire under a roof in lightweight construction can lead to rapid roof collapse. Vertical ventilation operations in lightweight roof trusses can slow roof collapse by removing heat and fire gases from the space below. Skylights can provide a quick and easy vertical ventilation point. In multi-story structures, there is the possibility of light or air shafts near the center of the structure, creating both a life hazard and an extension/exposure problem. You may not be able to identify the existence of light or air shafts from ground level. You must be aware of parapet walls and live loads such as billboards, air handling units, or water tanks on top of roofs because of their contribution to collapse potential **(Figure 3.18)**.

- ***Flat*** — Flat roofs are the easiest on which to work. Generally, they are the easiest to cut open and ventilate. Passive ventilation systems, skylights, monitors, and stairwell doors can be used for ventilation. At the same time, HVAC units and multiple layers of roofing material can increase the collapse potential. Steel decking (**Q-deck**, S-deck, or cm-deck [corrugated metal deck]) is used as a roof cover to support concrete flooring. It creates a hazardous situation if you start cutting panels and there are no supports under the area in which you are working. In addition, in some major metropolitan areas, steel plates have been used over flat roofs to prevent illegal entry. These plates prevent ventilation and create a collapse hazard. Parapet walls may also create a collapse hazard when exposed to fire or heavy caliber fire streams. Flat roofs may also conceal cocklofts that can spread smoke and fire well beyond the point of origin.

> **Q-Deck** — Also known as S-deck and corrugated steel deck. Structural steel decks used to support concrete flooring and roof membranes. May have fire retardant material applied to the underside of the structure.

- ***Pitched*** — A pitched roof is any roof that has a slope of 10° from horizontal of each side to the ridgeline. The pitch is generally stated in the number of inches of rise compared to 12 inches of horizontal run or length from the supporting wall to the ridgeline. For instance, a 12/12 pitch would have an angle of 45°. Pitched roofs are most often found on residential single-family dwellings **(Figure 3.19)**. Shape and pitch will determine roof operations, such as walking or using a roof ladder or an aerial ladder to gain access. In most peaked roof situations, the area under the roof will not give direct access to the living area. High peaks on residences can allow for the possibility of small apartments, bedrooms, and floored or unfloored storage space. Pitched roofs are generally ventilated near the ridgeline. However, opening the ceiling below may be impossible from this location without a long pike pole.

- ***Arched*** — Also known as a bowstring roof, the arched roof may have exposed rafters supporting the roof or be concealed by a drop ceiling. This type of roof was popular into the 1960s and used for grocery stores, warehouses, aircraft hangars, and theaters, among other types of occupancies. The use of parapet walls can conceal the presence of a bowstring roof from ground level. Bowstring roofs have a reputation for collapsing when exposed to fire **(Figure 3.20)**.

Figure 3.18 Smoke stacks and water tanks on roofs create collapse hazards in older buildings.

Figure 3.19 Some modern residential structures have metal or slate roofs that are set at an extreme angle creating a hazard to firefighters trying to work on them. *Courtesy of Ron Moore.*

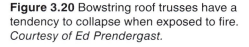

Figure 3.20 Bowstring roof trusses have a tendency to collapse when exposed to fire. *Courtesy of Ed Prendergast.*

- ***Rain roof*** — Generally pitched roofs placed over existing flat roofs. They may be constructed of wood trusses with plywood sheathing and composition or tile covering. Metal panels are also becoming popular for rain roofs. Rain roofs create a concealed space that can slow down or prohibit vertical ventilation and increase the potential for fire extension. HVAC units may be hidden from view under the rain roofs adding to the collapse potential. Rain roofs can hide fire spread, and the buildup of heated gases conceal possible explosive atmospheres.

Knowledge of the building size and configuration can offer the following information:

- ***General*** — The size of the building will be a strong factor in determining the number of resources needed and fire flow requirements. The larger the building, the more smoke and heat that can be absorbed by the building, hiding the fire location and extent. The quantity of smoke may not be an indication of the size of the fire in the structure.

- ***Area*** — The overall area or footprint of the building can provide an indication of the resources that may be required. Long distances from access points will dictate the length of hose or search lines needed. It will also indicate the time it will take search and attack crews to reach the fire or make a search while allowing time to withdraw. Air management becomes a priority as does finding possible closer and thus safer access points to the seat of the fire. The area will also determine the fire flow. A large area is a potential for a large fire requiring large fire flows **(Figure 3.21)**. Building codes may also require large-area structures to have standpipe connections at remote points as well as water-based fire protection systems.

- ***Ceiling height*** — High ceilings hide dangerous signs of rapid fire development. Hot fire gases can accumulate at the top causing rapid fire spread increasing the potential for collapse.

- ***Floors, below grade*** — Below grade floors may have interior or exterior access or both and/or windows used for ventilation and light. Reading the smoke on first floor can help you determine if the fire is below grade. If smoke does not rise when you open a door into the first floor area, then the fire is probably below you. Look for basement windows or smoke from the sill plate to identify signs of the basement fire. Observe the air movement

Figure 3.21 Modern multifamily residential structures can cover large areas and multiple floors. *Courtesy of Ron Moore.*

when the first floor door is opened. If smoke/air splits into layers with in-out movement in the doorway, the fire is probably on the first floor. Heavy smoke from eaves can signal a basement fire because of pipe chases reaching from the basement to the attic. This is a different situation from fire in a balloon-frame structure where fire can spread from floor to floor inside the walls. Finally, below grade fires are labor intensive requiring more resources to localize and extinguish them.

Single-Family Dwelling Fire Claims the Lives of Two Volunteer Firefighters — Ohio

On February 5, 1998, two male volunteer firefighters (Victim #1 and Victim #2) died of smoke inhalation while trying to exit the basement of a single-family dwelling after a backdraft occurred. A volunteer Engine company composed of four firefighters and one driver/operator were the first responders to a structure fire at a single-family dwelling 3 miles from the fire department. When the Engine company arrived, one firefighter on board reported light smoke showing from the roof. The four firefighters (including Victim #1) entered the dwelling through the kitchen door and proceeded down the basement stairs to determine the fire's origin. The four firefighters searched the basement which was filled with a light to moderate smoke. A few minutes later, a fifth firefighter from Rescue 211 (Victim #2) joined the group. After extinguishing a small fire in the ceiling area, Victim #2 raised a ceiling panel and a backdraft occurred in the concealed ceiling space. The pressure and fire from the backdraft knocked ceiling tiles onto the firefighters, who became disoriented and lost contact with each other and their hoseline. Two firefighters located on the basement staircase exited the dwelling with assistance from two firefighters who were attempting rescue. One firefighter was rescued through an exterior basement door and the two victims' SCBAs ran out of air while they were trying to escape. Both firefighters died of smoke inhalation and other injuries. Additional rescue attempts were made by other firefighters but failed due to excessive heat and smoke and lack of an established water supply.

Courtesy of NIOSH, Report 98-F06, June 16, 1998

- *Floors, above grade* — Floors above grade level may require ground ladder or aerial device access for search, rescue, and firefighter egress. Composite floor truss systems create multiple hazards including collapse potential when weakened by fire and under the weight of water used for fire extinguishment. The systems can also provide concealed avenues for fire spread. Windows and glass curtain walls will prevent access to upper floors. Watch for smoke extension on upper floors indicating possible fire extension within wall cavities. When smoke lifts or thins on the first floor but does not plane, or does not show any movement, the fire is probably located on an upper level.

- *Attic, cocklofts, concealed areas* — Fires in attics, cocklofts, and concealed areas can spread unseen and may smolder causing light smoke in a living area. Heat can cause build up of flammable gases resulting in flashover, smoke explosion, or rapid fire spread/growth. Roof trusses can be weakened by prolonged exposure to fire in these spaces.

You can learn a lot from the building access shown on the preincident survey or visible to you during your approach and 360° survey:

- *Street access* — Your first assessment is of the street access as you approach the incident scene. By being aware of what is going on in the neighborhoods in your response area, you should be aware of the general conditions. Look for closed streets caused by construction or occasional block parties that will delay arrival time. Be aware of parking patterns due to alternating parking, snow removal ordinances, or day/night parking requirements. Parking patterns may limit apparatus placement requiring responding units to use alternate routes. Upon arrival, engine officers should always consider aerial apparatus access and spotting locations on all streets

Figure 3.22 Narrow streets and congested parking areas can make aerial ladder operations difficult. *Courtesy of Ron Jeffers.*

but particularly narrow streets. Aerial apparatus can be completely useless if access is not provided. Loss of this asset can significantly change your attack plan. Aerial apparatus placement may limit street access or require using sidewalks for outrigger placement **(Figure 3.22)**.

- *Structure access* — On commercial occupancies, multiple locks, security grilles, and roll up doors delay entry, but are a good indication that the space is unoccupied. Residential occupancies may also have multiple locks and security grilles that will delay entry. A multiple family apartment dwelling may have numerous interior doors that need to be forced requiring use of a hydraulic ram/rabbit tool and/or assignment of additional members to forcible entry. Upon arrival, begin by determining how many access doors there are. In town house, garden-style, or apartment type buildings there may be only one entrance. In row houses or attached structures, members will be delayed from accessing the rear of the structure because of the distance they will have to travel.

- *Egress* — Firefighter safety depends on the ability to exit a structure rapidly if conditions change. Means of egress that you should consider are:
  - Doors: Note location of all doors and direction of swing. Determine if doors can be opened from the exterior. Note construction of doors and door frames in the event forcible entry must be used.

— Windows: Determine if windows can be opened. Find out the size of the window and determine if it is large enough for a person to fit through. Notice if windows are covered with security grilles or bars that will make egress difficult or impossible. Determine how high the windows are and if ladders will be needed to reach them. Solid nonopening windows are energy efficient but prevent access or egress and may allow heat and smoke to build up in the compartment or structure. Small windows between floors usually represent stair placement. Bathroom windows are usually wavy or bubbled for privacy and smaller than other windows.

— Stairs: Determine if exit stairwells provide egress directly to the outside of the building. Ensure that the exit doors can be opened from the exterior. Notice if stairwells permit reentry onto all floors from the stairwell. Find the location of all standpipe connections in the stairwells. Look for indications that the stairwells are pressurized or designed to remove smoke from the structure.

— Fire escapes: Consider the location, operation, and condition of any exterior fire escapes. Notice if the fire escape has been pulled down, indicating that it has been used by occupants to escape the structure.

— Dead-end corridors: Based on the preincident plan, consider the presence of dead-end corridors. These may exist in altered multifamily structures and small office buildings that use modular office units (cubicles).

- *Metal security bars* — Metal security bars make entry and egress difficult. Security on windows and doors may consist of metal bars, Plexiglas, wire grilles, or plywood or metal sheets. Steel skin doors or heavy metal grilles may be found on residences and commercial structures. Security doors, drop bars, and additional security are often found on the interior or exterior of rear doors to commercial occupancies. Always consider calling additional resources for heavily secured buildings, because forcible entry can be very labor intensive and exhaust crews rapidly. If the security bars are installed to code in residential occupancies, there should be an interior release on the bars for emergency egress.

Special hazardous conditions are rarely visible during your size-up. It is important that your preincident plan contain any information regarding these hazards which can tell you:

- *Alterations* — Changes to the interior can create multiple rooms with dead-end corridors, multiple doors, and no designated exit ways. Alterations can also create concealed spaces and disrupt the original coverage of sprinkler systems.

- *Change in occupancy or use* — Can create situations that the existing fire protection system cannot control. Corridor and stairwell walls, compartment doors, and means of egress may be insufficient for the new use. Occupancy load limits may exceed the ability of exit passages to carry the new load.

- *Storage* — Illegal storage, compressed gases, flammable liquids and explosive materials, or drug labs may also exist that will increase the life safety risk to firefighters.

Life Hazard

The life hazard to occupants will be largely determined by the occupancy type and the time of day. To be on the safe side, all structures should be considered occupied until proven otherwise. However, you must balance the risk to yourself and your personnel against the possibility that there is someone in the structure. ***Once you enter the structure, it is occupied and you are at risk.***

General Considerations

The following are general considerations that you must remember when determining the life hazard of various types of occupancies:

- In occupancies with large open areas, such as car dealerships, warehouses, factories, or places of assembly, assume that there are exposed trusses supporting the roof structure. These trusses may degrade rapidly when exposed to heat and fire creating a collapse hazard.

- Occupancy type will help with the rescue profile and estimating the potential fuel load. Assembly occupancies, such as nightclubs and theaters, have severe life safety dangers. Studies of the behavioral patterns of people indicate that most people will exit through the door they entered. However, building codes only require it to handle 50% of the occupant load to exit through that opening, providing there are other exits spaced evenly around the structure.

- Older commercial occupancies, such as warehouses or factories, may have been converted to residential occupancies creating additional problems of access, egress, and fire spread. At the same time, residential structures on main streets may have been converted to light office use, such as small medical clinics, dental offices, and law offices. These buildings may have multiple doors with deadbolt locks inside.

- Type of occupancy use can indicate the mobility of the occupants and their ability to self-evacuate. For instance, day care centers for young children and babies, group homes for the mentally and physically challenged, nursing homes, and senior care facilities will contain occupants who will require assistance in evacuating the facility.

- The name or type of business can indicate the possible contents and potential fire hazards to expect. Paint, hardware, pool supply, and automotive supply stores may contain flammable and combustible liquids; high thermal output materials, such as hydrocarbons and polymers; and caustic or hazardous materials. Rapid fire growth and high thermal output will cause you to consider exposures and evacuations earlier than normal.

Knowing the life safety hazards to firefighters is your primary concern. This knowledge includes:

- ***Air management*** — The IC must evaluate the size and complexity of the structure's interior to determine the amount of time required to exit the structure from the *point of no return*. The point of no return is the distance between the point work is to be performed and the nearest safe atmosphere. This calculation will determine the maximum amount of time that can be spent working in the IDLH. A general rule of thumb for calculating air supply is the capacity of the air cylinder divided by 2 equals the allowable working time. This calculation allows for a safety margin for exiting the structure.

- *Rehabilitation* — Although this is not an initial concern, you will need to watch and monitor your personnel. You should apply the two bottle rule to determine when to send personnel to rehab. This rule is based on 30-minute capacity air bottles or cylinders. You must refer to your department's SOP/SOG for local rehab requirements **(Figure 3.23)**.

- *Crew resources* — Hopefully a good size-up will determine your staffing needs early. Keep extra crews available, ensure that enough crews are on scene to clean up, and do not release resources too early. The IC should keep enough resources in staging to complete the incident. If you need to request additional resources, call for enough to mitigate the incident.

Figure 3.23 Local policy and procedures will determine the number of air cylinders a firefighter may use before being sent to rehab.

Residential

There are many types of residential occupancies. Single-family dwellings and apartments may be occupied 24 hours a day or just at night and on weekends. Hotels, motels, dormitories, and boarding houses are likely to be occupied 24 hours a day with staff on hand as well as guests or residents. Hotels, motels, and guest accommodations are likely to be occupied by transient residents who are not familiar with the structure, the exits, or the evacuation plan. Assistance in evacuating the structure may be required from the staff or from firefighters.

Multifamily dwellings may have unit access directly from the exterior or from a central interior hallway. These arrangements can affect your ability to advance hoselines and may increase reflex time. The number of doors, doorbells, mailboxes, and utility meters indicate the number of living units. Utility meters may indicate the number of units plus one for a building **(Figures 3.24a&b)**. Some apartment complexes, however, do not have separate utility service for every apartment. Exterior stairs with access to various levels can indicate multiple living areas. Window air conditioners, blinds, or curtains in windows may identify that the space is used as a living area.

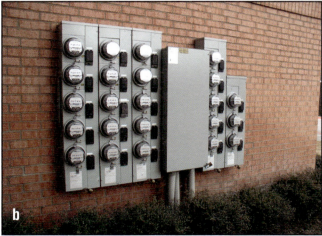

Figures 3.24a&b An indication of the number of individual units in a structure are the number of (a) gas meters and (b) electric meters. *Photo (b) Courtesy of Ron Moore.*

Mercantile and Business

Office buildings and retail commercial structures, such as grocery stores and lumberyards, are generally occupied only during business hours. Observe the normal operating hours and look for vehicles in front of the building that can indicate someone is inside. However, some retail outlets are open 24 hours a day while others may have staff inside the closed facility restocking shelves and cleaning the store. Likewise, some offices are staffed around the clock if they are involved in global enterprises or computer services. In some areas, small businesses may have sleeping quarters in the structure, whether or not their existence is code compliant.

Industrial

Depending on the type of manufacturing, industrial sites may be staffed around the clock. Generally, you will find security staff on duty at the site.

Institutional

Figure 3.25 When day care centers are in operation, staff should be present to assist in the evacuation of occupants.

Institutional occupancies include hospitals, nursing homes, day care centers, and detention/correctional institutions. Occupants may be classified as either ambulatory or nonambulatory. Ambulatory occupants are those who are capable of moving on their own, and nonambulatory persons are those who require assistance **(Figure 3.25)**. In detention, correctional, and psychiatric institutions, such as prisons, jails, and drug rehabilitation facilities, occupants may be confined or restrained. Institutional occupancies operate on a 24-hour basis. In many of these situations, protection of occupants in place is the more viable solution to life safety.

Assembly

Places of assembly vary in size and type. They include, among others:
- Restaurants
- Bars
- Theaters for live performances or movies
- Arenas
- Dance halls or ballrooms
- Nightclubs
- Places of worship

Places of assembly may operate 24 hours a day depending on the particular occupancy. Typically, exits are only required to be unlocked during business hours, but many types of assembly occupancies will have cleanup crews inside well after being closed to the public. To adhere to the occupant code (based on the building code), the owner/occupant must post the maximum number of people permitted in the room or structure. During business hours, exits may not be locked or blocked although exits may not be required to open from the outside of the structure, which can make firefighter entry difficult.

Educational

Educational occupancies, such as public and private schools, and technical schools are rarely occupied between midnight and 6:00 a.m. by students or faculty. It is not unusual, however for college and university libraries to be open after midnight. Security and cleaning staff may be present after the facility is closed. Like transient residents, students may not be familiar with the structure or exits unless the school has provided emergency training to them.

Unoccupied, Vacant, or Abandoned Structures

According to the NFPA®, an estimated average of 31,000 fires occurred in vacant structures between 2003 to 2006. At the same time, 15 firefighters died in vacant structure fires between 1998 and 2007. Annually 4,500 firefighters are injured in these types of fires. Every type of structure may be classified as unoccupied, vacant, or abandoned at sometime during its existence. Although these classifications are used interchangeably, they do have specific definitions:

- *Unoccupied* — A residential or commercial property that is for sale may be considered unoccupied when the occupant has moved out of it. The utilities will usually still be turned on and some contents may remain in the structure. The term may also be applied to a structure that is unoccupied when the business located there has closed. Fire protection systems may be operational.

- *Vacant* — A residential or commercial property that is empty and may have all entrances secured or boarded up. Vacant structures may still be on the local property tax roles. The owners may be contributing to the structure's upkeep, and it may still have value. The utilities may or may not be turned on and some contents may still remain in the structure. Fire protection systems may or may not be operational.

- *Abandoned* — A property that has been vacant for some time. It may be structurally unsound, or it may contain conditions that are in violation of the local building code. There may also be some question as to legal ownership of the structure. The utilities will have been terminated due to lack of payment. Fire protection systems may not be operational. Although the terms *vacant* and *abandoned* may be used to describe the same conditions, the key difference is the length of time the structure has been unused or uninhabitable.

Some jurisdictions, for example New York City, have the authority to inspect and designate abandoned or vacant structures as *Vacant*. According to local ordinances, the owners must secure the structure to prevent access and post signs indicating the building is vacant. In those cases, fire department SOP prohibits interior fire attacks.

Vacant or Abandoned Structures

Signs that a vacant or abandoned structure may be occupied include:
- Electrical extension cords leading into the building
- Candles or campfires burning
- Evidence of forced entry
- Secured doors or windows that have been opened
- Make-shift curtains/shades or decorations
- Signs of construction or demolition
- Presence of Dumpsters®, construction material, parked vehicles, and temporary electric or utility meters

When making your size-up, it is important to implement a risk/benefit analysis for all buildings that are suspected to be unoccupied, vacant, or abandoned. Too many firefighters are being injured or killed in the growing population of vacant buildings. Recent reports by the IAFF and the CDC both provide insights into the hazards of these types of structures. Hazards that may be found in vacant or abandoned buildings include:

- Delay in discovery and notification of fire
- Fires located in multiple locations
- Increased exposure hazard
- Open shafts, pits, or holes in floors
- Structural instability caused by weather, vandalism, or lack of maintenance
- Exposed structural members
- Penetrations in fire barriers that would permit rapid fire extension
- Mazelike configuration in interior floor plan
- Blocked or damaged stairs
- Combustible contents

Resources

Your size-up will also include a review of your resources starting with those that are immediately available to you and extending to those that are available on call. Your immediate resources are your crew, apparatus, and equipment as well as the crews, apparatus, and equipment that were dispatched for the alarm. If you have any doubt regarding what resources are responding, ask the dispatcher to clarify. Other resources that you have available are the fire protection systems at the site and the water supply, which must be considered in your evaluation of the situation.

Your Apparatus Resources

Your apparatus will provide you with your initial water supply, tools, equipment, and hose to attack or contain the fire. This resource is of limited value if you do not have the personnel to perform the various duties for which the equipment is designed. Nowhere is this more evident than in the volunteer departments where one person may drive the apparatus to the fire and wait for other volunteers to arrive in their own vehicles. In career departments, minimum staffing generally ensures that a crew arrives on scene with enough people to perform at least some of the duties.

Fire Protection Systems

The preincident plan will indicate the existence of on-site fire protection systems including water-based suppression systems (sprinklers), non-water-based suppression systems (dry chemical, foam, etc), and standpipe systems. The location of the fire department connection (FDC) and the suppression system control valves should be clearly marked on the plan. It is possible, however, that the system is out of service, damaged, or inadequate to control the fire or contain it to one area. You must be prepared for the possibility that you may not be able to rely on the system. The preincident plan should indicate the location of standpipe outlets in multistory and large area structures and the need for any adapter fittings to convert the system to fire department hose couplings.

Knowing about the fire protection system will allow you to know to look for:

- **General** — An activation alarm, such as a water-motor gong or electronic alarm, is an indication the system has been activated. Depending on the type of monitoring system, the dispatch center may be able to tell you the zone within the building that has been activated. Otherwise, you should access the system control room or panel. Determine the location and means for shutting off or silencing all fire protection systems. Your local SOP/SOG will dictate the process for shutting off systems and returning them to service. Sprinkler systems should not be shut off until you are in a position to control the remaining hazards.

- **Water-based fire protection system** — Besides the activation alarm, water may be visibly flowing from the system drain or from the interior of the structure. Be aware that some systems may only protect a small area of the structure such as exit passageways, storage rooms, or concealed spaces **(Figure 3.26)**.

Figure 3.26 Fire suppression systems may have multiple control valves within the system.

- **Non-water-based fire protection system** — A special system may have activated that should completely extinguish the fire. Such systems include a range hood system or a clean agent Halon-type system in a computer room or electrical space.

- **Standpipe system** — You should know the type of system, wet or dry, the location of the FDC, and the location of standpipe outlets and control valve. Knowing the location of the outlets will help determine the length of hose required to reach the most remote area of the fire floor.

- **Detection and alarm systems** — Depending on the type of monitoring system, the dispatch center may be able to tell you the zone within the building that has been activated. Otherwise, you should access the system control room or panel.

Water Supply

Your initial visual size-up, your preincident plan, and your experience should tell you if your apparatus water supply is sufficient to control the fire. **Table 3.1** provides a general guide for estimating water supply based on tank capacity. At the same time, you must consider the probability that you will need a continuous water supply provided from a municipal water system, a private water supply, or from static water supplies. The preincident plan should indicate the location of each of these sources. Hydrant spacing, which will determine the length of supply hoseline that you will need, is generally constant in most urban areas and some rural areas. However, the condition of hydrants may make them unserviceable or inadequate to supply the amount of water you will need. You must be prepared to assign units to perform relay pumping or request water tenders to shuttle water to the site.

Your ability to control a fire will depend mainly on proper tactics and the amount of water that you have available. The preincident plan or your knowledge of the area will give you a basic idea of the available water sources. Ask yourself, "Is the size of the fire such that I can achieve knockdown with my apparatus water tank? How much fire growth can occur before I can get water on the fire? Would my tank supply be better suited to protect exposures and write off the original structure if there is no life hazard?" If you are going to need to depend on a water tender or water shuttle, you must have an idea of how long it will take to set up the water supply.

Table 3.1
Estimating Water Supply Based on Tank Capacity

Tank Capacity	Flow Rate	Approximate Time
500 gallon (2 000 L)	150 GPM (600 L/min) 250 GPM (1 000 L/min)	3 minutes 2 minutes
1000 gallon (4 000 L)	150 GPM (600 L/min) 250 GPM (1 000 L/min)	6 minutes 4 minutes
1500 gallon (6 000 L)	150 GPM (600 L/min) 250 GPM (1 000 L/min)	10 minutes 6 minutes
2500 gallon (10 000 L)	150 GPM (600 L/min) 250 GPM (1 000 L/min)	16 minutes 10 minutes
3000 gallon (12 000 L)	150 GPM (600 L/min) 250 GPM (1 000 L/min)	20 minutes 12 minutes

Other Available Resources

Your initial assignment may not include all the types of apparatus or equipment that you will need to control the fire. In particular, you may need to request water tenders to provide a continuous source of water. Specialized apparatus such as hose tenders with large diameter hose (LDH) or compressed air foam (CAF) units may be available by request. In some jurisdictions, aerial devices and breathing air supply apparatus may not be included on the initial assignment **(Figure 3.27)**. Finally, additional personnel may need to be called to the scene arriving in their personal vehicles and assigned as they arrive.

Figure 3.27 Some departments may have air supply apparatus equipped with spare air cylinders and fill stations.

Arrival Condition Indicators

Often initial decisions must be made with incomplete information. When you make your initial decisions, you may not know all the facts or have the information you need. You may have to depend on probabilities and indicators as well. The time of day, current and projected weather conditions, and visual indicators will all help you to make sound decisions. Then, you make your initial fire attack or rescue if you have the necessary resources to do so safely.

Time of Day

As mentioned in the Occupancy section, the time of day can indicate whether the structure is occupied and if the occupants are awake or asleep. The structure still needs to be searched to ensure that all occupants are out. Time of day will also give you an idea of the amount of traffic congestion adjacent to the structure and the possibility that on-street parking will obstruct direct access. For instance, the street in front of most public schools is crowded with waiting school busses and personal vehicles around the time school is dismissed. Visibility will be limited during the hours of darkness requiring the use of artificial lighting and reflective barrier to protect the street side of apparatus **(Figure 3.28)**. Fires in occupancies that are closed may have a delayed transmission of the alarm resulting in an advanced fire.

Figure 3.28 Limited visibility at night or in heavy smoke conditions can increase risks to firefighters. *Courtesy of Chief Chris Mickal, New Orleans Fire Department.*

Weather

Weather conditions can have a negative effect on structural fire fighting operations by reducing the effectiveness and efficiency of your resources. Besides slowing your response to the scene, ice, snow, rain, humidity, temperature extremes, and wind can make it difficult for personnel to work. In addition, occupants who are removed from the structure will have to be moved to a place of shelter.

- *Ice* — Ice is not always visible, creating a slipping hazard. It makes advancing attack or supply hoselines, performing vertical ventilation, and doing forcible entry difficult. Hoses, ladders, and apparatus can become ice covered causing slipping hazards.

- *Snow* — Snow creates some of the same difficulties as ice. It also obscures tripping hazards under its surface.

- *Rain* — Rain and fog can reduce your ability to see the entire scene. It can make metal surfaces slippery and freeze on equipment and apparatus as the temperature drops.

- *Humidity* — High humidity can cause smoke to remain close to the ground obscuring visibility of the building. It can also affect personnel causing them to tire quickly and become dehydrated through perspiration loss.

- *Temperature extremes* — In extremely hot climates, personnel may succumb to heat stress rapidly and require rehabilitation earlier than normal. They may also become dehydrated requiring additional fluids and medical care. Extreme cold temperatures can cause skin and clothing to stick to metal tools and equipment, cause frostbite injuries, and reduce stamina.

Hoselines, pumps, and water supplies can freeze causing a loss in water supply or pressure. Hose, tools, and equipment can be damaged or become inoperable.

- **Wind** — In windy conditions, flame spread can become dangerously rapid if windows fail or are opened improperly. Strong winds can cause working on the peak of a roof to become very dangerous. A fire that appears to have started inside a structure may be the result of a wind-driven fire that began on the outside and extended into the structure. It is important to determine the source of the fire and not make an interior attack that is not needed initially. High velocity winds can spread fire brands onto exposures and can affect the accuracy and effectiveness of defensive hoseline streams.

Visual Indicators

Visual indicators provide you with insight on conditions upon which to provide the initial report and base the initial decisions. Visual condition indicators vary widely depending on the type of emergency incident you encounter. A large plume of black smoke may indicate a hydrocarbon fire with accelerants or a structure containing a large quantity of plastic materials. Your 360° survey is very important because the large plume of black smoke may be an outside fire in the rear of the structure. You should never rely on only one indicator to make a decision **(Figure 3.29)**. Once a decision is made, you must continually size up the situation looking for indicators to confirm that your decision was correct. If indicators are contradictory to your decision, you must reevaluate and change the plan.

The size, location, and extent of the fire may not be obvious to you during your initial size-up. The presence of a cockloft may permit smoke to be present far from the point of origin. Multiple basement levels may require added time searching each level before the seat of the fire is located. Basements may become flooded by broken water lines, activated fire protection systems, or runoff from attack lines creating additional hazards. Also, unprotected wood or composite I-beams that are exposed to fires in basements and may fail within 3-5 minutes.

Various indicators, including smoke, heat, and flame will assist in determining the size, location, and extent of the fire. The following Information Box provides some structural fire condition indicators.

Figure 3.29 Thick smoke may be caused by multiple factors including the type of fuel or stage of the fire.

Structural Fire Condition Indicators

The following indicators of structural fire conditions based on the appearance of smoke, air track, heat, and flame may be visible to the first-arriving personnel:

- **Smoke** — Visible products of combustion. Tar, soot, and carbon are the most common heated particles found in smoke giving it the black color. Water moisture and heated gases give the smoke its white color. The large particles cool quickly. Because of their size, they drop out or are filtered out of the smoke, changing the black smoke to gray then white.
  Indicators:

  — Light white indicates that pyrolysis (chemical change by heat) is occurring in areas adjacent to the main body of fire. The light white color indicates moisture and gases are being released from the product. White smoke has lost particles because the particles have cooled from travel, water, and the addition of very cold air. In the early stages of a fire, white wispy smoke is moisture coming off objects. A deep seated fire where the smoke has been filtered or cooled from travel will also be white **(Figure 3.30)**. On very cold days when the temperature is below freezing, smoke turns white and turbulent immediately on leaving the structure. When smoke is forced from a structure through cracks, it filters large particles and will be white.

  — White smoke explosion is rare. However, when a product is heated at a consistent temperature, it can pyrolyze and release types of flammable gases. When they accumulate in the right mixture, they become explosive when they come in contact with an ignition source.

  — Thick, white smoke is smoke that was thick and black but has lost the solid particulates that made it black. This can happen when smoke has been cooled to very cold temperatures or has travelled a long distance inside a structure while losing heat. Its consistency is fluffy as compared to a steam cloud which is more opaque and mostly consists of condensation or water mist.

  — Brown smoke is common in mid-stage heating as moisture mixes with gases and carbon as pyrolysis increases. It is also very common in mid- to late-stage heating to see the caramel color brown smoke. Brown smoke is an indication of unfinished wood burning. Caramel-colored smoke usually indicates clean wood burning such as a fire involving structural wood members.

Figure 3.30 White smoke may have been filtered or may be an indication that water is being applied to the fire.

continued

- Gray smoke indicates a combination of mixing. It can be mid-stage heating with white, brown, or black, or it can be when different smoke areas combine. It can indicate smoke production changes from mid-stage heating to very high heat.

- Black smoke contains high quantities of carbon particles and is also an indicator of the amount of ventilation available at the seat of fire. The thicker the smoke, the less clean burning and the less oxygen available, as smoldering fires produce massive amounts of black smoke. In the past, hydrocarbon fires were considered the source of black smoke. Now, fires involving synthetics, plastic, resins, polymers and products made from hydrocarbon derivatives will give off large quantities of black smoke. This black smoke contains unburned fuels and is a good indicator of carbon monoxide and many other flammable gases.

- Thin, black smoke is the direct result of heat from a flame. This smoke is black, but possible to see through. Thin, black, fast-moving active smoke is an indication the fire is nearby. Thin, black smoke with smooth lines exiting high in an opening and going straight up indicates flame-driven smoke, meaning open flames are nearby with good ventilation.

- Thick, black smoke indicates the late stages of pyrolysis, which produces large amounts of carbon as unburned product. This indicates high carbon monoxide gas in the smoke, creating a highly flammable condition. The term *black fire* refers to dark, black, and thick turbulent smoke (fuel) that is ready to ignite. Vent point ignition is a possibility. This area would not be survivable by an occupant if the smoke existed floor to ceiling. This thick black smoke can act like flame, cause pyrolysis, and can even char. Black fire is very hot and can be seen traveling quickly out openings in hallways and other channels. Once the ventilation plane is unable to feed the large growing fire, it will become ventilation controlled and fill the area becoming a backdraft or smoke explosion hazard. Generally, water is applied to active fire only. With the presence of black fire, applying water to cool the ceiling area is encouraged in an effort to reduce the potential for a flashover.

- Unusual color smoke should give the IC an indication that different extinguishing agents may be needed. Flammable metals and chemicals will give off uncharacteristic colors of smoke as they burn.

- Location of smoke may be an indicator of the location of the fire or a false indicator caused by movement of smoke through a structure. The color and velocity or pressure can indicate if smoke has traveled distances from the origin.

- Height of the neutral plane (separation between the under pressure [lighter air] and overpressure [heavier air] regions of a compartment): Also known as thermal balance; the neutral plane lowers as a fire develops and density of fire gases (smoke) increases. Interior attack teams must observe the level of the neutral plane very carefully and prepare to withdraw when there is a rapid change in conditions causing the neutral plane to lower. Levels:
    - High neutral plane: May indicate that the fire is in the early stages of development. Watch the danger of high ceilings; they can hide the dangers of a fire that is in later stages. A high neutral plane can also indicate a fire above your level.

continued

- Mid-level neutral plane: Could indicate that the compartment has not ventilated yet and that flashover is approaching.
- Very low-level neutral plane: May indicate that the fire is reaching backdraft conditions. This could also mean a fire is below you.

- **Volume** — Quantity of smoke visible. Small buildings will fill up with smoke sooner than larger buildings. However, a large building pushing white smoke from openings is a sign of a large fire. Smoke must fill the building. As it travels it will lose its black color and become volume pushed, not very active. Lots of black smoke from a small house can be burnt food displaying a large quantity of smoke from front door on arrival. Volume-pushed smoke will usually flow, neither smooth nor turbulent. It floats out of openings, rising slowly.

- **Density** —The darker and more turbulent the smoke is, the closer you are to a rapid fire event. Indicative of a very large fire or an underventilated fire is lots of suspended particles of tar, carbon, ash, and soot. Thick black smoke has lots of heat; thick, white smoke has traveled and lost particles but still may have plenty of heat and fuel to burn. The denser the smoke, the lower the visibility, and the more likely that heat buildup indicates a pending flashover. Fast-moving black smoke can create fast moving fires with rapid fire spread (black fire).

  An important point to remember is that smoke is a combustible by-product of a fire and will burn rapidly when exposed to enough heat.

- **Air track** —Movement of fresh air toward the base of a fire and movement of smoke and heated air out of a compartment; understanding this phenomenon can help in the ventilation of a fire. Indicators:

  — Velocity and direction: Slow, smooth movement of air toward a fire indicates that it is in the early stages and still fuel-controlled. Air movement is rapid and turbulent when a fire becomes ventilation-controlled. A sudden and total rush of fresh air into a compartment can indicate that a backdraft condition is imminent. The direction of airflow can indicate the location of the seat or base of the fire.

  — Pulsations: Fuel-rich and oxygen-deficient conditions result in smoke pulsing out of openings in a closed structure. Opening the structure improperly or at the wrong location can rapidly add fresh oxygen to the compartment, resulting in a backdraft or move the fire to an undesired location. With these conditions, ventilation operations should be limited to above the fire and attack should not be initiated until ventilation has been accomplished.

  — Noise: Whistling noise created by the movement of the air into the structure indicates that a backdraft condition may be imminent.

- **Smoke Movement** — Three ways can describe how smoke moves: floating or hanging, volume pushed, and heat pushed. The heat in the smoke will dictate speed.

  — Floating or hanging (lost its heat) smoke is the same temperature as the air around it. This is often found in air conditioned buildings, fires that are sprinkler controlled, or where smoke particles are filtered and cooled by passing through cracks in walls. Floating or hanging smoke will move according to air currents. Lazy smoke indicates a small, early stage fire, mostly containing moisture from the first stage of pyrolysis. It could also mean a deep-seated fire in a large building but the smoke is cooling off as it travels great distances.

continued

concluded

- — Volume-pushed. Once the area fills up with smoke, the speed it pushes out is important. The faster it exits, the more the conditions produced indicate a large fire and the speed of possible growth. Slower smoke exits in slow rolls that are not turbulent or active.

  When smoke fills the compartment, it becomes very pressurized. It can reduce flame action if underventilated. If confined it can be seen forced out of cracks around windows, doors, and eves. The color and volume can indicate room location and contents or structure fire. The area from which the smoke comes will indicate passage for smoke, which means passage for fire. Black smoke means the fire has an open area of travel, while white indicates a filtered area of travel.

  — Heat pushed is indicated by speed, which can be turbulent (bubbling, boiling, active) or laminar (smooth straight). Fast, angry, turbulent, or active smoke means there is a serious working fire. Turbulent black smoke has lots of particles and, indicative of vent-controlled smoke, has heat. It pushes from openings coming out rolling up and to sides as more smoke pushes out into it, bubbling, rolling, active or angry. Laminar, thin, fast-moving, straight-line smoke is heat pushed, (not underventilated) and is exiting from openings near the active flaming fire.

- *Heat* is a form of energy transferred from one body to another as a result of a temperature difference. Indicators:

  — Blackened or crazed (patterns of short cracks) *glass:* are an indicator of fire in the room or nearby as hot smoke condenses on a cooler window. While crazing indicates high interior temperatures, this is an indicator mostly seen on single pane windows. Take caution when opening a structure with these indicators.

  — Blistered paint: Indicates both temperature extreme and location of the neutral plane. It may also indicate fire behind the wall.

  — Sudden heat buildup gives a very late indicator of flashover. When operating inside a structure, personnel must be aware of a rapid increase in temperature and act immediately to apply water or exit the structure.

- *Flame color* — Color is usually an indicator of the oxygen supply and the extent of fuel-oxygen pre-mixing, which determines the rate of combustion.

Summary

Successfully controlling any emergency incident will depend on your ability to make informed and accurate decisions. Your decisions are based on the proper use of the size-up process. The size-up process begins with a preincident survey of the property. During the preincident survey, you become familiar with the target hazards within your response area. Although most departments only have resources to survey high life safety or high value sites, some are able to perform surveys of many or all commercial structures in the community. Informal surveys can be made of residential properties during voluntary home inspections or medical responses. An important part of the preincident survey is to determine the fire flow requirements for the target hazards and for the average commercial and residential structures. The size-up process continues through the emergency response as you combine the preincident survey with the information provided during the dispatch to the incident scene. Your initial decisions will be based on everything you know plus what you see upon arrival. Your size-up will continue as you deploy your resources, select the appropriate strategy, and assign tactics and tasks to control the incident. It has been said that the decisions made by the initial IC in the first five minutes can determine the outcome of the next five hours of the operation.

Review Questions

1. What is size-up?
2. How is size-up theory applied preincident, on arrival, and during the incident?
3. Why are preincident surveys conducted?
4. What information should be gathered during a preincident survey?
5. What facts should be gathered during a preincident survey?
6. What is included in an arrival report?
7. What items does Layman describe that are needed for analyzing any emergency situation?
8. What are the critical fireground factors needed for size-up?
9. How can knowledge of roof types and styles affect size-up?
10. What general guidelines should be considered when determining the life hazard of various types of occupancies?

Strategy

Chapter Contents

Incident Priorities 141
    Life Safety 142
    Incident Stabilization 142
    Property Conservation 142

Risk vs. Benefit 143

Operational Strategies 144
    Offensive Strategy 145
    Defensive Strategy 145

Managing the Incident Scene 148
    Develop Incident Action Plan 148
    Resource Tracking...................................... 149
    Implementing the IAP 152

Lloyd Layman's Tactical Model 155

Summary 156

Review Questions 156

Courtesy of Chief Chris Mickal, New Orleans (LA) Fire Department.

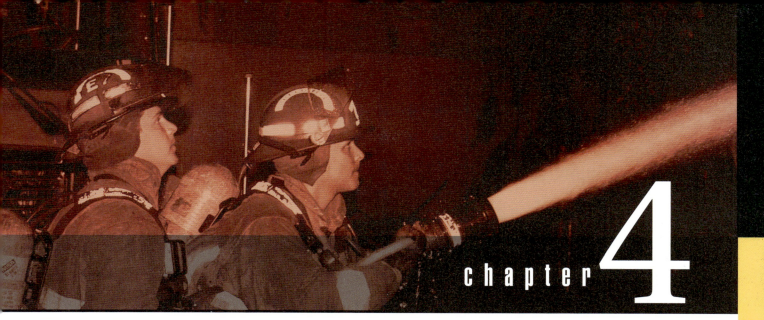

chapter 4

Key Terms

Fireground...142
Freelancing 149
Line of Duty Death (LODD) 143
Mayday...146
Personnel Accountability
   Report (PAR) 147
RECEO/VS Model 155

Strategy

Learning Objectives

After reading this chapter, students will be able to:

1. Describe each of the incident priorities.
2. Discuss risk vs. benefit.
3. Describe each of the operational strategies.
4. Explain managing the incident scene.
5. Describe Lloyd Layman's tactical model.

Chapter 4
Strategy

Your initial incident scene size-up results in the development of the incident priorities. The simplest acronym to remember the priorities for structural fire strategies is LIP:

- Life safety
- Incident stabilization
- Property conservation

To achieve your incident priorities, you must select an appropriate strategy. When determining the strategy to use, you must take into consideration the level of risk you are willing to accept. You must weigh the potential injury or loss to your resources against the value of the lives or property that you can save. This procedure is known as a risk-benefit analysis.

As discussed previously, some departments create preincident plans from preincident surveys for specific types of hazards. These plans can be the foundation of the Incident Action Plan (IAP) that you will use to control various types of situations. Preincident plans help to determine the strategy and tactical activities of units that respond to the incident site. If your department uses preincident plans, remember that each incident may be different. Therefore all, part, or none of the preincident plans may be applicable to the given situation.

Finally, your strategy should lead to attainable tactical objectives. To assist in selecting the proper tactics in an efficient manner, you need a model that is easy to remember and apply. The model used in this manual is the one developed by Chief Lloyd Layman, which has been successfully used in the fire service for many years.

Selecting the appropriate incident priorities, developing strategic goals, and learning the tactical model of Chief Layman are the topics covered in this chapter. This chapter provides a foundation for the in-depth discussion of tactics in Chapter 5.

Incident Priorities

Life safety, incident stabilization, and property conservation (LIP) are the incident priorities that apply to all types of incidents and listed in order of priority. By placing them in this order, it is easier to make strategic and tactical decisions.

Life Safety

Life safety is always the first priority and exists when a structure is known or suspected to be occupied. The reason for this should be obvious: your first concern is the life safety of your crew, other responders, and yourself. Attempting to save a vacant building or to attempt to rescue victims who are beyond saving is not worth risking the life of a firefighter. Firefighter life safety is based on a number of criteria, including but not limited to:

- Using the correct personal protective equipment (PPE) properly **(Figure 4.1)**
- Assigning tasks based on the abilities of the personnel
- Having appropriate training
- Applying a risk/benefit assessment model
- Following the local standard operating procedures or guidelines (SOP/SOG) for fireground operations
- Following industry approved tactics
- Maintaining discipline on the **fireground**
- Using tools and equipment correctly

The second concern of life safety is locating, stabilizing, protecting, and removing living victims from the hazardous area. If you know that someone is still alive and still savable in the structure, use an offensive strategy to rescue them. If there is no reason to believe that there are survivable victims in the structure, adopt the strategy appropriate to stabilize the incident and conserve property with every consideration for firefighter safety.

Incident Stabilization

Incident stabilization is the process of controlling the fire. While incident stabilization is listed as a lower priority than life safety, it may be necessary to implement it first as a means of providing life safety. For example, it may be necessary to place attack lines into operation and/or initiate ventilation in order to halt fire progress thereby providing a safer entry for search teams. Incident stabilization can also result in property conservation through a rapid fire attack that prevents the extension of a fire into unburned compartments or areas of the structure(s). Rapid fire attack may also result in reduced water damage due to the use of fewer gallons (liters) for extinguishment. Incident stabilization is not achieved until the fire is under control **(Figure 4.2)**.

Property Conservation

Because most property can be replaced or may be protected by property owner's insurance, the goal of property conservation is the last incident priority. Traditionally, property conservation is thought of as salvage or loss control activities. Like incident stabilization, some property conservation activities may begin early in the incident and be performed simultaneously with the other two priorities. For instance, exposure protection should always be paramount once life safety has been addressed **(Figure 4.3)**.

While the conservation of property has to be viewed as a necessary activity, at no time should property conservation take precedence over life safety or incident stabilization. If a defensive strategy is implemented, then strict adherence to collapse zones must be followed for firefighter safety.

Fireground — Area around a fire and occupied by fire fighting forces.

Figure 4.1 Firefighter life safety includes the proper use of personal protective equipment which is determined by the type of hazard that is present.

Figure 4.2 The incident is considered stabilized when the fire is under control. *Courtesy of Bob Esposito.*

Figure 4.3 The use of waterproof covers to protect property has long been an essential element of property conservation.

Risk vs. Benefit

As a fire and emergency service responder, you are expected to take calculated risks to provide for life safety, incident stabilization, and property conservation. Calculated risks mean that you do not blindly go into a situation, but rather that you gather information through size-up and determine what level of risk is acceptable for the given situation. For the past two decades, the fire service has continued to have the same number and types of **line-of-duty deaths**. Therfore, in 2001 the International Association of Fire Chiefs (IAFC) developed a model policy called *The 10 Rules of Engagement for Structural Fire Fighting*. The policy was developed to help ensure that everyone can return home safely.

Line-of-Duty Death (LODD) — Firefighter or emergency responder death resulting from the performance of fire department duties.

> ### The Ten Rules of Engagement for Structural Fire Fighting
>
> **Acceptability of Risk**
>
> 1. No building or property is worth the life of a firefighter.
> 2. All interior fire fighting involves an inherent risk.
> 3. Some risk is acceptable in a measured and controlled manner.
> 4. No level of risk is acceptable where there is no potential to save lives or savable property.
> 5. Firefighters shall not be committed to interior offensive fire fighting operations in abandoned or derelict buildings that are known to be or reasonably believed to be unoccupied.
>
> **Risk Assessment**
>
> 1. All feasible measures shall be taken to limit or avoid risks through risk assessment by a qualified officer.
> 2. It is the responsibility of the Incident Commander to evaluate the level of risk in every situation.
> 3. Risk assessment is a continuous process for the entire duration of each incident.
> 4. If conditions change, and risk increases, change strategy and tactics.
> 5. No building or property is worth the life of a firefighter.

Your decisions are made on acceptable risk based on the initial assessment of the incident scene. This concept is clearly stated in a decision-making model developed by the Phoenix (AZ) Fire Department (PFD). The model is a departmental standard operating procedure (SOP) that is used to help PFD officers make reliable emergency response decisions. The essence of the model is stated as follows:

- Each emergency response is begun with the assumption that *responders can protect lives and property*.
- Responders will *risk their lives a lot, if necessary, to save savable lives*.
- Responders will *risk their lives a little, and in a calculated manner, to save savable property*.
- Responders will *NOT risk their lives at all to save lives and property that have already been lost*.

Firefighter safety is your primary concern. Your decisions must be based on this model which should be applied to your initial size-up. When committing to an interior offensive attack, you must balance the risk to your personnel if there is a doubt that a savable life is at risk.

Operational Strategies

Once you have determined the appropriate level of risk, you must decide on the strategy to implement. Traditionally, there are two strategies used by the fire service. You must select one of these strategies: *offensive* or *defensive* **(Figures 4.4 a & b)**.

Figures 4.4a&b (a) Offensive operations generally involve advancing hoselines into a structure (b) while defensive operations are conducted from outside the structure. *Photo (b) courtesy of Chief Chris Mickal, New Orleans (LA) Fire Department.*

Offensive Strategy

The offensive strategy used at a structure fire usually means that resources are deployed for interior operations to accomplish incident priorities. Countless possible variations exist for this scenario, depending upon the following conditions:

- Life hazards
- Structural stability

By talking to building occupants who have escaped the structure, neighbors, or other witnesses, the Incident Commander (IC) may be able to determine whether there are any occupants still inside and, if so, whether there is a reasonable chance that any of them are still alive. When taking into account the two factors listed above, an offensive strategy may be justified. When resources permit, fire attack would be started simultaneously with a search and rescue operation. If not, the search might have to be delayed until the fire is confined (stabilized) or a hoseline is placed between the occupants and the fire.

The offensive strategy may involve rescue or extinguishment or both. In some fire incidents, rescue and extinguishment will occur simultaneously with engine crews attacking the seat of the fire, while other personnel search for victims. In extreme cases, where a victim is known to be trapped, rescue will become the *primary* activity and fire attack will be performed only to protect the rescuers and the victim.

Defensive Strategy

The decision to operate in a defensive strategy is indicated when no threat to occupant life exists, occupants are not savable, or when the property is not salvageable. The defensive strategy is intended to isolate or stabilize an incident and keep it from getting any worse or larger. In the case of a structure fire, a defensive strategy may mean sacrificing a building that is on fire to save adjacent buildings that are not burning. A defensive strategy is usually (but not always) an exterior operation that is chosen because of insufficient resources or an interior attack is unsafe.

Defensive strategic operations involve personnel and apparatus that are kept at a safe distance from the hazards of the incident. This strategy may be employed if the incident is too large or hazardous to safely resolve with an offensive strategy (such as large structure fires or potential structural collapses). For example, protecting exposures from the spread of a large fire rather than extinguishment becomes the primary mission of responders. Defensive strategic operations are justified in the following conditions involving a structure fire:

- *Volume of fire* — Amount of fire exceeds the resources available to confine or extinguish. Examples include:
  - Lack of personnel or trained personnel.
  - Inability to provide adequate fire flow; that is, gallons per minute (gpm) (liters per minute [L/min]) because of insufficient pumping capacity or availability of water supply
  - Lack of appropriate apparatus or equipment to implement the required tactics

- *Structural deterioration* — Structure is unsafe for interior entry.
- *Risk outweighs benefit* — If the amount of risk to emergency responders is greater than the benefit.
- *Structural condition* — Structure is known to be vacant or abandoned.

If the initial IC determines that a defensive strategy is justified, resources may be assigned to apply water to protect adjacent exposures and/or other action to prevent the fire from spreading to these exposures. A defensive strategy may also be employed if resources are limited to the point where offensive operations cannot be implemented safely.

In an escalating incident such as a working fire, the IC may be forced to use an exterior defensive attack in the early stages because of resource limitations or the volume of fire may require the use of heavy streams directed from outside the structure. But as additional resources arrive at the scene, it may be possible, depending on the effectiveness of exterior attack streams, to switch from a defensive to an offensive strategy.

When firefighter fatalities and injuries occur at structural fires, they are sometimes the result of firefighters being in offensive positions on defensive fires or rapid changes in the situation such as the following:

- Structural member weakening can cause a ceiling, floor, wall, or other structural member to collapse.
- Introduction of fresh air into a superheated compartment can cause a rapid fire development condition (backdraft, flashover, etc.) **(Figure 4.5)**.
- Heavy or dense smoke can obscure vision causing the firefighter to become disoriented and run out of breathing air.
- When a firefighter is trapped or unaccounted for and a **Mayday** has been sounded.
- Lack of a proper Incident Command System in place and failure to change strategy when the situation changes.

> **WARNING!**
> Defensive and offensive strategies should **NEVER** be attempted concurrently in the same area of the structure. Exposure protection can be accomplished as a property protection strategy and does not always mean defensive operations are the strategy.

Mayday — International distress signal broadcast by voice.

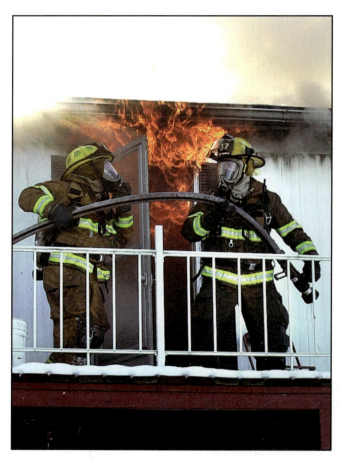

Figure 4.5 Extreme fire conditions can occur without warning before hoselines are charged or in position. *Courtesy of Bob Esposito.*

- IC not gathering all available information to complete a proper size-up (360° view of the structure).
- Wind-driven fire changes conditions quickly.

These changes may occur and the IC may not be aware of them. It is important for Incident Commanders to conduct **personnel accountability reports (PAR)** to check on the welfare of all firefighters on the scene every 10 to 20 minutes, or in accordance with your department's SOP/SOG, or when any one of the above conditions has occurred.

When a situation rapidly changes, the IC needs to be prepared to transition the operation from an offensive to a defensive strategy. The change requires the IC to communicate the change to all personnel and units operating at the incident. A PAR should follow to ensure that all personnel have been advised and have withdrawn from offensive positions. This change in transition is not up for debate. Many times the interior or roof crew will try to sway the IC's decision. Although the crew members may feel that they can extinguish the fire or complete vertical ventilation, the crew does not have the same information and exterior view of changes that the IC can see.

How the transition occurs depends on the speed at which the situation changes. All personnel must be made aware of the transition. Some units may need to remain in place to protect the withdrawal of other units. Supervisors/company officers must always know the location of personnel assigned under their command and must conduct personnel accountability checks when withdrawal is complete. During an orderly withdrawal, hoselines should *not* be abandoned unless absolutely necessary, as opposed to an emergency evacuation in which case all personnel evacuate with all due haste. Abandoned hoselines cannot provide any protection during a withdrawal. RIT/RIC personnel must be ready to assist any units that require assistance during the transition. Companies should continue to perform their assignments during the transition.

For a transition to be efficient and effective, it must occur as soon as the need is recognized. The IC must maintain current and accurate knowledge of the situation. The IC must be aware of changing conditions throughout the structure. Officers must give situation or status reports to the IC or their immediate supervisor regularly.

> **Personnel Accountability Report (PAR)** — A roll call of all units (crews, teams, groups, companies, sectors) assigned to an incident. Usually by radio, the supervisor of each unit reports the status of the personnel within the unit at that time. A PAR may be required by SOP at specific intervals during an incident, or may be requested at any time by the IC or the ISO.

Prince William County (MD) Department of Fire and Rescue
Line-of-Duty Death (LODD) Report for Technician I Kyle Robert Wilson

On April 16, 2007, the Prince William County Department of Fire and Rescue experienced its first line-of-duty death in the department's 41-year history. Technician Kyle Robert Wilson, a 15-month member of the department, died in the line of duty while performing search and rescue operations at a house fire in the Woodbridge area of Prince William County, Maryland. On that day, Technician Wilson was part of the firefighter staffing on Tower 512, which responded to the house fire that was dispatched at 0603 hours. The Prince William County area was under a high wind advisory as a nor'easter storm moved through the area. Sustained winds of 25 mph with gusts up to 48 mph were prevalent in the area at the time of the fire dispatch to Marsh Overlook Drive.

continued

Initial arriving units reported heavy fire on the exterior of two sides of the single-family house, and crews suspected that the occupants were still inside the house sleeping because of the early morning hour. A search of the upstairs bedroom commenced for the possible victims. A rapid and catastrophic change of fire and smoke conditions occurred in the interior of the house within minutes of Tower 512's crew entering the structure. Technician Wilson became trapped and was unable to locate an immediate exit out of the hostile environment. Mayday radio transmissions were made by crews and by Technician Kyle Wilson of the life-threatening situation. Valiant and repeated rescue attempts to locate and remove Technician Wilson were made by the fire fighting crews during extreme fire, heat, and smoke conditions. Firefighters were forced from the structure as the house began to collapse on them and intense fire, heat, and smoke conditions developed. Technician Wilson succumbed to the fire, and the cause of death was reported by the medical examiner to be thermal and inhalation injuries.

The major factors in the line-of-duty death of Technician I Wilson were determined to be:

- Size of the initial arriving fire suppression force
- Size-up of fire development and spread
- Impact of high winds on fire development and spread
- Large structure size and lightweight construction and materials
- Rapid intervention and firefighter rescue efforts
- Incident control and management

The weather conditions and construction features resulted in the rapid and catastrophic progression of fire conditions. The organizational preparation and response to incidents of this nature can and are recommended to be improved with the majority of recommendations focused on staffing, training, procedures, and communications.

Source: Prince William County Department of Fire and Rescue LODD Report. For additional information, see Appendix C for the NIOSH LODD Report.

Managing the Incident Scene

Depending on local policy and procedures, the first arriving fire officer or acting fire officer is responsible for managing the incident scene. When you are in that position, you must act in accordance with the locally adopted ICS. Besides determining the incident priorities and strategy previously mentioned, you must also develop and implement the Incident Action Plan (IAP). This plan will allow you to assign resources based on the desired results, track your resources, and accomplish the tactics required to fulfill the incident priorities.

Develop Incident Action Plan

The Incident Action Plan (IAP) is based on the information you gathered in the incident size-up. The IAP may or may not result in a written plan. If the incident is small and can be handled by the first-alarm assignments, the plan does not have to be written down. As the incident grows in size, the use of a "tactical work sheet" can be of great value to the IC. Examples of Tactical Work Sheets can be found in Appendix D. If the incident is large or has the potential for involving multiple units or agencies for a prolonged period of time, it must be in writing. Standardized ICS forms are available to record the various elements of the plan.

The plan must be communicated to all units and individuals operating at the scene before they are given a work assignment. This communication is done in person or over designated radio frequencies. All incident personnel must function within the scope of the IAP; actions taken outside the scope of the IAP are called **freelancing** and may place responders in jeopardy and reduce operational effectiveness. Fire officers should follow standard operating procedures identified by the agency and/or the IC. Incident personnel should direct their actions toward achieving the incident objectives, strategies, and tactics specified in the plan. When all members understand their positions, roles, and functions in the ICS, the system can safely, effectively, and efficiently use resources to accomplish the plan.

Freelancing — Operating independently of the IC's command and control.

Resource Tracking

One of the most important functions of an ICS is to provide a means of tracking all personnel and equipment assigned to the incident. In most career departments, units responding to an incident arrive fully staffed and ready to be assigned an operational objective. In volunteer and combination departments, personnel may arrive individually or in small groups. They will have to be formed into working units before they are committed to the incident.

Fire departments should have in place SOP/SOGs to guide personnel in establishing a tracking and accountability system at incidents. The SOP/SOGs must incorporate the following elements which must be included in the IAP:

- Procedure for checking in at the scene **(Figure 4.6)**.
- Method or plan of identifying the location or assignments of each unit and all personnel on scene.

> **CAUTION!**
> Personnel who arrive individually at an incident must always check in at the Command Post or Staging Area if it has been implemented. Personnel must not *freelance* or assign themselves to a task. Freelancing is unsafe and has resulted in fatalities and injuries.

Not necessarily part of the IAP, procedures for releasing resources no longer needed should be included in department SOP/SOGs. Personnel must be trained to check out when leaving the incident.

The IC must be able to locate, contact, deploy, and reassign the units that are assigned to an emergency incident. These tasks are accomplished through the ICS procedures that assign units to locations within the operating area. As units arrive on the scene, the IC assigns them to locations or functions as needed. The units may be held in a staging area until needed or until they are released from the incident. If staging has not been implemented, company officers must check in with the IC at the Command Post for an assignment upon their arrival.

Figure 4.6 Tracking and accountability systems include a procedure for personnel to check in before entering the incident scene.

Communications between units will be through the jurisdiction's emergency radio communication system or through direct face-to-face communication **(Figure 4.7, p. 150)**. As units are assigned to the incident, the central communications center may announce the command frequency in use or automatically place all radios on that frequency. Units that have been assigned to the incident must ensure that they have complete communication with the Command Post. Face-to-face communication is always the preferred method of exchanging information, but is not always practical on the fireground. There are some communications that should be done over the air so that all personnel on the fireground are aware of the information.

The IC can use a number of visual aids to help manage and track resources assigned to the incident. The visual aid can be as simple as a preprinted form, called *Tactical Work Sheets* **(Figure 4.8)**, to sketch the incident scene and the location of units as they arrive, by using white boards with grease pencil or Velcro® for tracking accountability tags or passports. Another visual aid that can be used is an elaborate tracking board with magnetic symbols identifying each of the units. Regardless of the tracking device, it should be simple to read and contain as much information as necessary about the activities of all the units on scene. The visual aid may contain the following information:

Figure 4.7 The preferred method for making unit or personnel assignments is through face-to-face communication.

- Assigned radio frequencies
- Assigned, available on-scene, and requested units
- Activated ICS functions
- Site plan
- Staging areas
- Logistics location
- Control zones
- Unit incident assignment and location

Each organization must develop or adopt a system of accountability that identifies and tracks all personnel working at an incident. The organization should standardize the system so that it is used at every incident. If the system is used on the small, routine incidents, it will be more likely to work on large, complex incidents. All personnel must be familiar with the system and participate in the system when operating at an emergency incident. The system must also account for those individuals who respond to the scene in vehicles other than emergency response apparatus, including staff vehicles

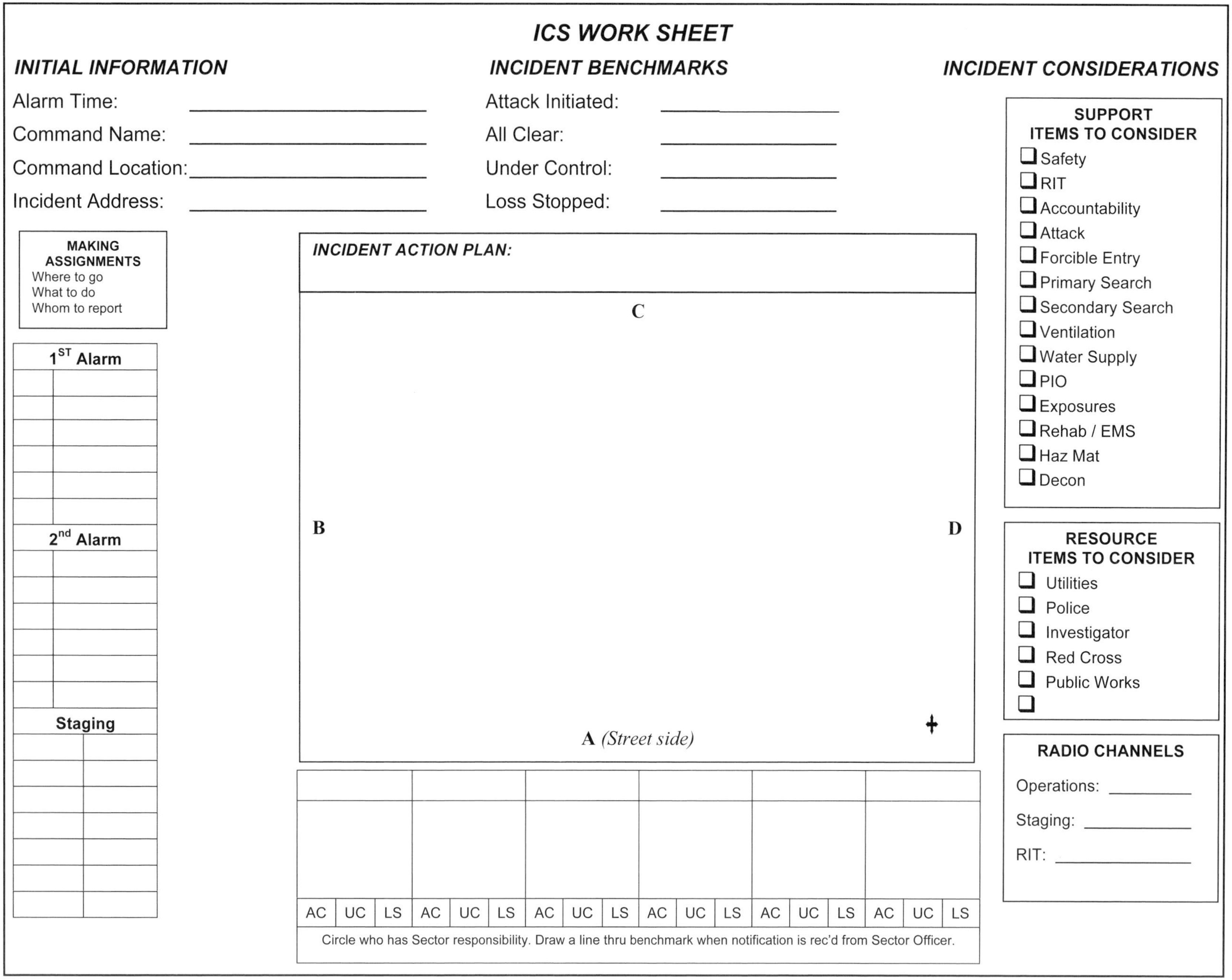

ICS WORK SHEET

INITIAL INFORMATION

Alarm Time: _____
Command Name: _____
Command Location: _____
Incident Address: _____

MAKING ASSIGNMENTS
Where to go
What to do
Whom to report

1ˢᵀ Alarm

2ⁿᵈ Alarm

Staging

INCIDENT BENCHMARKS

Attack Initiated: _____
All Clear: _____
Under Control: _____
Loss Stopped: _____

INCIDENT ACTION PLAN:

C

B   D

A (*Street side*)

AC	UC	LS	AC	UC	LS	AC	UC	LS	AC	UC	LS	AC	UC	LS	AC	UC	LS

Circle who has Sector responsibility. Draw a line thru benchmark when notification is rec'd from Sector Officer.

INCIDENT CONSIDERATIONS

SUPPORT
ITEMS TO CONSIDER
☐ Safety
☐ RIT
☐ Accountability
☐ Attack
☐ Forcible Entry
☐ Primary Search
☐ Secondary Search
☐ Ventilation
☐ Water Supply
☐ PIO
☐ Exposures
☐ Rehab / EMS
☐ Haz Mat
☐ Decon

RESOURCE
ITEMS TO CONSIDER
☐ Utilities
☐ Police
☐ Investigator
☐ Red Cross
☐ Public Works
☐

RADIO CHANNELS
Operations: _____
Staging: _____
RIT: _____

Figure 4.8 Example of a Tactical Work Sheet.

and personally owned vehicles (POVs). In multiagency operations, officers must ensure that the personnel accountability systems of the different agencies are coordinated at the scene.

Accountability is vital in the event of a change in the condition of the incident. At a structure fire, that change might be the extension of the fire through a concealed space or the rapid increase in the volume of fire creating a flashover or backdraft situation. The IC must know who is at the incident and where each person is located. For example, self-contained breathing apparatus (SCBA) can malfunction or run out of air, and firefighters can get lost in mazes of rooms and corridors. Without having an accountability system, it is impossible to determine who and how many may be trapped inside or injured. Too many firefighters have died because they were not discovered missing until it was too late. A well-run accountability system with a Command Officer requiring constant feedback in the form of personnel accountability reports (PAR) are two very important components to tracking and maintaining firefighter safety.

Personnel operating in the IDLH must be assigned in crews of two or more and be close enough to each other to provide immediate assistance in case of need. Crew members must be in constant communication with each other through visual, audible, or physical means and have a portable radio to keep in contact with the IC. Company officers are responsible for keeping track of the members of their units.

Where no life safety hazard is present, the initial interior attack should not be attempted until there are sufficient personnel on scene to provide one crew inside and another crew outside acting as a standby crew, available for immediate rescue of crew members requiring such help inside. See NFPA® 1500 for further details. As discussed in the previous chapter, it is important to ensure that personnel assigned to standby for rescue of crews operating inside are not performing other tasks that, if abandoned for crew rescue, would further endanger any other personnel on the scene.

The IC is responsible for managing the personnel accountability system employed by the organization. The IC may delegate the authority by assigning an accountability or Incident Safety Officer to monitor the system. The system should indicate the individuals assigned to each apparatus, the names of people responding individually, such as staff personnel and volunteers, the time of arrival, the assigned duty or unit, and the time of release from the scene. Various systems, such as the tag system, SCBA tag system, passports, electronic GPS systems, and barcode readers are available for tracking individuals at the emergency incident **(Figure 4.9)**.

Figure 4.9 Accountability tags may be placed in the apparatus that personnel are assigned to for a particular operation.

Implementing the IAP

Putting your IAP into action is the next step in the process of controlling the emergency incident. At this point, you must select a command option and allocate your resources. Your choice of command options will be based on how hazardous the incident is, how rapidly it may develop, and how many resources are currently available to you. In some departments, you may have all the personnel, apparatus, and water supply within minutes of your arrival. In others, you may have to wait for a long time before you can deploy your resources.

Command Options

With the priorities established and your IAP ready for use, you must be ready to take Command of the situation. As the first-arriving officer, you have the following three Command options available:

- *Investigation Option (Nothing-showing)* — When the problem generating the response is not obvious to you, you should establish Command and announce that *nothing is showing*. Then direct the other responding units to stage at a location which would allow maximum flexibility for their deployment of their tactical assignment at the incident based on the local SOP/SOG. You then accompany unit personnel on an investigation of the situation and maintain Command using a portable radio. This approach applies to all types of most emergencies.

> **Caution**
> It is essential that you are prepared to provide a continuous water supply to the incident.

- *Fast-Attack or Mobile Command Option* — When your direct involvement is necessary for the Unit to take immediate action to save a life or stabilize a marginal situation (an instance where the officer must remain with his/her crew), you should take Command and announce that the Unit is initiating a *fast attack*. You enter the structure (IDLH) with your crew and direct the attack on the fire. You will continue in the fast-attack option, which usually lasts only a short time, until one of the following situations occurs:

  — Incident is stabilized.
  — Incident is not stabilized, but you must withdraw to outside the hazardous area to establish a formal Incident Command Post (ICP).
  — Command is transferred to a superior officer, or the initial IC is reassigned to another function.

  NOTE: Under no circumstances shall anyone be in an IDLH without a partner. Crews must remain together with no fewer than two personnel.

- *Command Option (Command Post Option)* — Because of the nature, scope, or potential for rapid expansion of some incidents, immediate and strong overall Command is needed. In these incidents, you should assume Command by naming the incident and designating the location of the ICP, giving an initial report on conditions, requesting the additional resources needed, and initiating the use of a tactical work sheet. With this option, the company officer remains at the mobile radio in the apparatus, assigning tasks to other unit personnel, communicating with other responding units, and expanding the NIMS-ICS as needed by the complexity of the incident (**Figure 4.10**). You may assign one of the other members of your

Figure 4.10 The IC may establish a Command Post in the cab of an apparatus located at the incident scene.

crew as the acting officer, or direct the other members of your crew to work under the supervision of another company officer or assign them to other ICS positions. Before doing so, remember that crews assigned to another commander are no longer available to you, and you must ensure that their absence will not create a hardship.

When you need to transfer command of an incident to another officer, the transfer must be done correctly. Otherwise, there could be confusion about who is really in Command of an incident. The officer assuming Command must communicate with you by radio or through face-to-face communication, which is preferred. *Command should never be transferred to anyone who is not on the scene.* Command should only be transferred to improve it, that is, to a higher ranking officer, a more experienced team member, or to get the initial IC back to his or her crew for close supervision. When transferring Command, you should brief the relieving officer on the following items:

- Incident status (such as fire conditions, number of victims, etc.)
- Safety considerations
- Goals and objectives listed in the Incident Action Plan (IAP) or tactical work sheet
- Progress toward completion of tactical objectives
- Deployment of assigned resources
- Assessment of the need for additional resources

NOTE: Command CANNOT be transferred or passed to an officer who is NOT on the scene

Resource Allocation

As the IC, you must be able to determine if the available resources can support the initial or sustained attack. If the available fire flow and resource capabilities are equal to or greater than the estimated need, an initial offensive attack can be made. If the fire flow requirement exceeds the available water supply and/or resource capabilities, a defensive attack must be implemented. In situations where there is no life hazard and the initial offensive attack is unsuccessful in controlling the fire or if the fire increases in intensity, the IC must increase the fire flow applied to the fire, with additional larger hoselines. If this is not possible then you must shift to a defensive strategy until additional resources become available or the incident is terminated. All personnel and units as well as the communication center must be notified of the change of operational strategy, and attacking units must be given time to disengage in an orderly manner. Personnel accountability is essential for a safe and efficient transition to the defensive strategy.

NOTE: If the available fire flow is equal to or greater than the estimated need, an initial offensive attack can be made. If the available fire flow is less than the estimated need, an initial defensive attack must be made.

Your next action is to assign your resources, which can be further divided into two steps: Assigning members of your Unit and assigning other Units that have been or will be dispatched to the incident.

1. Your Personnel — This may be the more difficult of the two actions especially if you have minimal staffing. Dividing up your personnel to perform the various tasks that make up the most critical tactics and still adhere to local SOPs and legal guidelines can be very difficult. Generally, one person will be assigned to remain with the apparatus, usually the driver/operator. Limited staffing may not permit you to remain at the Command Post when you are needed to investigate the source of the fire.

Your greatest challenge will be deciding the level of risk at which to place your personnel when a life safety situation is obvious or perceived **(Figure 4.11)**. The two-in, two-out rule can effectively prohibit you from making entry into the IDLH environment.

2. Other Units — During your initial size-up, you must consider the number and type of units that are responding with you and the length of time until they are ready to deploy at the scene. If a preincident plan (operational plan) exists for the site, then your job of assigning subsequent units will be somewhat easier. However, you must not depend on it completely. The type of situation you encounter may be very different from the one that the plan is intended to address.

Figure 4.11 When resources are limited, assigning personnel can be very challenging.

If no preincident plan exists or the situation is unique, assign arriving units based on your tactical needs to meet the strategic goals and department SOP/SOG. A key to a successful operation is to assign units so that apparatus do not have to be repositioned during the incident. Assigning units to a staging area allows you time to determine the best location and use of your personnel and equipment.

Lloyd Layman's Tactical Model

Fire and emergency services personnel respond to a wide variety of types of emergencies. In every emergency situation, whether it involves a fire, hazardous materials release, rescue, vehicle extrication, or medical response, the incident priorities are basically the same: life safety, incident stabilization, and property conservation.

If your department does not have SOP/SOGs in place to help in the decision-making process when responding to structure fires, then you should apply Chief Lloyd Layman's decision-making model referred to by its acronym **RECEO/VS**. Layman recognized the need for identifying priorities in emergency situations. Even though his list of priorities is stated in fire-control terms, he also acknowledged that the same priorities could be applied to any type of emergency. Chief Layman's model, which he developed in the 1940s, was intended as a way to train fire officers and firefighters in a method of critical thinking. He recognized that learning from experience at emergency incidents took a long time and was inefficient. Instead, he proposed to teach the model in training classes and scenarios.

RECEO/VS Model — One of many models for prioritizing activities at an emergency incident: Rescue, Exposures, Confine, Extinguish, Overhaul, Ventilation, and Salvage.

Initially, Layman intended the term *RECEO* to be both a list of priorities *and* a sequence of operations. He did not include *ventilation* and *salvage* in the list because they are not needed in every fire incident and not always performed at the same point in the fire incidents. Current application of the model includes ventilation and salvage as part of the process that may be inserted any place they are needed. The initials stand for the following priorities or operations but not necessarily in priority order:

- **Rescue** — Identifies the life-safety aspect of emergency incident priorities
- **Exposures** — Describes the need to limit an emergency incident to the property or area of origin
- **Confinement** — Describes the need to confine an emergency incident to the smallest possible area within the property of origin
- **Extinguishment** — Describes the activities needed to resolve an emergency incident
- **Overhaul** — Describes activities that restore an incident scene to a condition that is as nearly normal as possible while checking for extension
- **Ventilation** — Describes activities that control or modify the environment
- **Salvage** — Describes all actions taken to protect structures and contents from preventable damage

While some documents refer to these priorities as strategies, Chief Layman referred to them as tactics. A complete description along with application of each of these tactics is included in the following chapter. You should make a habit of mentally applying these to each structural fire you are assigned to and practice making decisions based upon this list.

Summary

To successfully contain, control, and extinguish a structural fire, you must determine the incident priorities based on life safety, incident stabilization, or property conservation. In some situations, these priorities can be accomplished at the same time, in support of each other. Your next step is to select and apply a strategy, either offensive or defensive. Your choice of strategies will be based on the highest incident priority, the level of risk to your resources, and the resources you have available. The strategy is met by developing an Incident Action Plan, which is based on an operational plan. The strategy is also met by assigning resources to accomplish the variety of tactics as defined by Chief Lloyd Layman's model. These tactics are discussed in the following chapter.

Review Questions

1. What are the incident priorities that apply to all types of incidents?
2. Upon what criteria is firefighter life safety based?
3. What are the major points of the risk vs. benefit model of the Phoenix (AZ) Fire Department?
4. What are the two operational strategies? Describe each briefly.
5. What is an Incident Action Plan (IAP)?

6. What Command options are available when implementing the IAP?
7. When transferring Command, what information should be provided to the relieving officer?
8. What does RECEO/VS stand for? Describe each priority briefly.

Tactics

Chapter Contents

Rescue **161**
    Search Safety Guidelines 164
    Conducting a Search 165

Exposures **168**
    Interior Exposures 169
    Exterior Exposures 169

Confinement **171**

Extinguishment **173**

Overhaul **175**

Ventilation **178**

Salvage **181**

Summary **183**

Review Questions **183**

Courtesy of Chief Chris Mickal, New Orleans (LA) Fire Department.

chapter 5

▌ Key Terms

Kerf Cut 172
Search Line 165
Trenching 172

Tactics

Learning Objectives

After reading this chapter, students will be able to:

1. Discuss rescue and rescue tactics.
2. Describe types of exposures.
3. Explain confinement.
4. Explain extinguishment tactics.
5. Describe overhaul operations.
6. Describe ventilation tactics.
7. Explain salvage operations.

Chapter 5
Tactics

In Chapter 1, you learned that tactics are a series of operations performed at the Division or Group level that, in combination, accomplish an overall strategy. Tactics are measurable and attainable outcomes that lead to the completion of the strategy and incident objectives.

In the model developed by Chief Lloyd Layman introduced in the previous chapter, tactics are prioritized based on life safety, incident stabilization, and property conservation. **Table 5.1** illustrates the relationship between the tactics and incident priorities.

Chief Layman first recorded his theories of fire suppression tactics in the book *Fundamentals of Fire Fighting Tactics* in 1940 when he was the chief of the Parkersburg (West Virginia) Fire Department. It was updated by the NFPA® in 1953 under the title *Fire Fighting Tactics*. In addition to these original documents, Chief Layman also did research and training in the use of fog stream nozzles during World War II. In 1952, NFPA® published Layman's book, *Attacking and Extinguishing Interior Fires*. This chapter dicusses each of the tactics Chief Layman used in his text.

Table 5.1
Tactical Objectives and Intended Incident Priorities

Tactic	Incident Priority
Rescue	Life Safety
Exposures	Life Safety Incident Stabilization Property Conservation
Confinement	Life Safety Incident Stabilization Property Conservation
Extinguishment	Life Safety Incident Stabilization Property Conservation
Overhaul	Incident Stabilization Property Conservation
Ventilation	Life Safety Incident Stabilization Property Conservation
Salvage	Property Conservation

Rescue

The tactic referred to by Chief Layman as *rescue* actually consists of two components: search and rescue. The search component requires that you, your crew, or other units physically locate any victims who are in the hazard area. The rescue component requires that once you locate the victims, you separate them from the hazard **(Figure 5.1, p. 162)**. Separating the victims from the hazard involves moving them outside the involved structure or moving them to an area of safety located inside the structure, referred to as *sheltering-in-place*. Rescue may also be accomplished by placing a charged hoseline between the victims and the fire.

Making the decision to perform search and rescue can be a very subjective process. It is based on facts that you know, perceptions that you believe to be true, and projections of what you think may occur in the future. Your ability to

Figure 5.1 Rescue is the act of removing a victim from a hazardous area.

complete a risk assessment, assign tasks, and determine a mode of attack is dependent on your understanding of the facts, perceptions, and projections — your situational awareness.

Search is a systematic approach to looking for possible victims. *Rescue* is the act of removing victims known to be in danger. Assuming that there may be people in the structure is a Command decision, and placing firefighters to do a search is not the same as a rescue. If there is a possibility that people may be in the structure, a primary search is conducted. The tactics in supporting a primary search then become as important if not more so when searching.

The need to perform search and rescue will be determined by facts, perceptions, and projections that you will start evaluating long before the alarm sounds. You may take the approach that a life will be in danger, and rescue will be required in all cases. This scenario will generate a thought pattern that includes an interior attack with search teams and ventilation. However, this scenario may also be beyond the ability of your resources, placing you and your crew at greater risk than necessary. While making the decision to commit to the rescue tactic, you must fully consider the facts available in developing your perceptions and projections. The plan should take into consideration where to start the search, where victims are most likely to be found, and the dangers involved. Fire attack and ventilation must be coordinated with the search effort. In some cases, confinement of the fire must be initiated prior to the search and rescue effort for the sake of the firefighters' and occupants' safety. Confinement of the fire may be necessary before committing personnel to rescue. Confinement options include creating a barrier between occupants and the fire, using ventilation to draw the smoke and fire away from occupants, or placing a hoseline in operation between you and the fire.

Searches are normally conducted by an entire crew being supplemented by an attack and ventilation crew. However, a minimum search crew would consist of two firefighters. The officer in charge of the search team will need to determine the resources required to perform the search. Size and complexity of the structure as well as the time of day will drive the decision for the number of personnel needed for search operations. The report of smoke in a hotel at 2 a.m. will require more search teams than a similar situation in a warehouse on a weekday at 2 p.m.

Tactics are achieved through the completion of tasks performed by units and individuals. Multiple tasks may be required to accomplish the tactic of rescue. Some of these tasks may be associated with other tactics, such as ventilation or fire extinguishment.

Rescue Tasks
- Performing size-up
- Performing ventilation
- Gaining access
  — Forcing entry
  — Setting ladders as needed
- Performing primary search
- Deploying charged hoselines
- Confining the fire
- Conforming to the two-in, two-out rules of your organization
- Locating victims
- Protecting victims from an IDLH environment
  — Shelter in place
  — Remove from hazard area
  — Separate victims from hazard

Generally, rescue will require you to use an offensive strategy with an interior attack supported by coordinated horizontal or vertical ventilation **(Figure 5.2)**. When the rescue is complete, reassess your strategy and risk management plan. Make tactical decisions that should concentrate on confinement and extinguishment utilizing either an offensive or defensive strategy.

Figure 5.2 Rescue teams must coordinate their actions with fire attack and ventilation teams.

Search Safety Guidelines

The following is a list of safety guidelines that search and rescue personnel should use when operating within a burning building:

- Do not enter a building in which the fire has progressed to the point where viable victims and property cannot be saved. Report to your supervisor any conditions that would appear to indicate there are no savable victims. This is a conscious decision you make concerning life and death. If you choose not to attempt a rescue, you must be able to justify your decision. You must also be able to justify the decision to attempt a rescue that results in casualties in your crew because adequate resources were not available.

- Observe if backdraft conditions are apparent. Entry should only be attempted after appropriate ventilation is accomplished.

- Work according to the Incident Action Plan (IAP). Do not freelance. If you need to deviate from the IAP, this information needs to be reported to the IC.

- Maintain radio contact with your supervisor (Command, Division or Group Supervisor, etc.), and monitor radio traffic for important information.

- Monitor continuously the fire conditions that might affect the safety of your search team and other firefighters.

- Use the established personnel accountability system without exception.

- Be aware of the secondary means of egress established for personnel involved in the search.

- Wear full personal protective equipment, including self-contained breathing apparatus (SCBA) and personal alert safety system (PASS) device (and be sure the PASS device is turned ON).

- Work in teams of two or more and stay in constant contact with each other. Rescuers are responsible for themselves and each other **(Figure 5.3)**.

- Search systematically to increase efficiency and to reduce the possibility of becoming disoriented.

- Stay low and move cautiously while searching where visibility is limited.

Figure 5.3 Search teams must operate in pairs remaining in contact with each other at all times.

- Stay alert and maintain situational awareness.
- Monitor continuously the structure's integrity and fire conditions, and communicate any significant change.
- Check doors for excessive heat before opening them.
- Chock open doors in the search area if this will not spread the fire.
- Mark entry doors into rooms and remember the direction turned when entering the room. To exit the building, turn in the opposite direction when leaving the room.
- Maintain physical contact with a wall, a hoseline, or a **search line** when visibility is obscured. Working together, search team members can extend their reach by using tether or branch lines. In a large area structure with signs of a working fire, even with good visibility, firefighters should *never* leave the hoseline or search line and must always remain in direct contact with the rest of the team.
- Have at hand charged hoseline whenever possible when working on the fire floor (or the floor immediately above or below the fire) because it may be used as a guide for egress as well as a tool for fire fighting.
- Coordinate with ventilation teams before opening windows to relieve heat and smoke during search.
- Use caution when conducting an interior search without a hoseline in the immediate area. If you encounter fire in a compartment, close the compartment door, and report the location and condition.
- Inform your supervisor immediately of any room or rooms that could not be searched, for whatever reason.
- Report promptly to the supervisor once the search is complete. Besides giving an "all clear" report (primary or secondary search complete), also report the progress of the fire and the condition of the building.
- Communicate hazard(s) found to the IC (for example, floor openings, heavy dead loads on the roof).
- Give clear directions for assignments and get feedback of understanding.
- Provide a safety brief with tactical assignments.

> **Search Line** — Consists of 200 feet (60 m) of 3/8-inch (10 mm) rope with a Kevlar® sheath to provide abrasion protection and heat resistance. Steel rings are located at 20 foot intervals (6 m) for the attachment of tether or branch lines used for performing lateral searches from the main search line.

Conducting a Search

In most structure fires, the search for life requires two types of searches: primary and secondary. A *primary search* is a rapid search that is performed either before or during fire suppression operations. The search should be started in the structure where occupants would be in the most danger. This area is normally above the fire and the rooms adjacent to the fire. It is often carried out under extremely adverse conditions. However, in situations where there is a high probability of the structure being occupied, it must be performed as soon as possible. In these situations a charged hoseline should be nearby.

During the primary search, check the known or likely locations of victims as rapidly as conditions allow, moving quickly to search all affected areas of the structure as soon as possible. Also during the primary search, the search team or teams can confirm that the fire conditions are as they appeared from the outside or report any changes they may encounter.

A *secondary search* is conducted *after* the fire is under control and the greatest hazards have been controlled. Ideally, when resources are available, the secondary search should be conducted by personnel other than those who conducted the primary search. The secondary search is a slower and more thorough search that attempts to ensure that no occupants were overlooked during the primary search. The secondary search team should approach the task as though the space or compartment has not been searched before.

Primary Search

Primary searches can be divided into two different types based on the type of occupancy or structure. In large area commercial buildings, the initial effort needs to be to search for and locate the seat of the fire and gain control. It is rare to see a civilian rescued from one of these buildings, which usually have detection and alarm systems and fire suppression systems that will alert occupants. In these buildings, victims are usually found while the entire crew is advancing hoselines.

In single-family dwellings and small commercial occupancies, it is possible to make searches quickly without a charged hoseline. Once resources are available, additional personnel must advance a hoseline into the structure to support and protect the search team.

Vent, Enter, and Search (VES)

VES is the acronym for vent, enter, and search. VES is a procedure used when fire has cut off the normal means of entry into or egress from part of the building. Credible reports indicate that one or more victims are trapped in a room cut off by the fire, but which can be accessed from outside. This technique may involve the use of ladders to access upper areas of a structure for 1) ventilation in front of the attack crew and 2) rapid, primary searches of individual rooms.

During the primary search, always use the buddy system — working in teams of two or more. By working together, two rescuers can conduct a search quickly while maintaining their own safety. When searching in an immediately dangerous to life and health (IDLH) atmosphere, you must remain in physical, voice, or visual contact with other team members.

Primary search teams must have at least one working portable radio, Thermal Imaging Camera (TIC), if available, flashlight/hand light, and forcible entry tools with them when they enter the building and throughout the search **(Figure 5.4)**. Valuable time will be lost if rescuers have to return to their apparatus to obtain basic tools and equipment. Also, tools used to force entry may be needed to force a way out of the building if rescuers become trapped.

Figure 5.4 The search team must be fully equipped with hand lights, forcible entry tools, portable radio, and Thermal Imaging Camera, if available.

When searching within a structure, the protection of a hoseline is recommended. This will provide a way for you to remain oriented so that you can find your way out quickly if fire conditions change rapidly during the search. Move systematically from room to room, searching each room completely, while constantly listening for sounds from victims. On the fire floor, start your search as close to the fire as possible and then work back toward the entrance door. This allows your team to reach those in the most danger first — those who would be overtaken by any fire extension that might occur while the rest of the search is in progress. Because those who are a greater distance from the fire are in less immediate danger, they can wait to be reached as your team moves back toward safety. To reach a point nearest the fire, proceed as directly as possible from the entry point.

When it is necessary to search large areas, a search line may have to be used. It will be your lifeline to safety and will help you cover the space completely and systematically. Always consider the use of a search line in large area structures, even with only light smoke showing, because conditions could change rapidly.

Secondary Search

After the initial fire suppression and ventilation operations have been completed, personnel other than those who conducted the primary search are assigned to conduct a secondary search of the building. During the secondary search, speed is not as important as thoroughness. The secondary search is conducted just as systematically as the primary search to ensure that no rooms or spaces are missed. But the secondary search is conducted more slowly and comprehensively than the primary search. Furniture is moved away from walls, floor-level cabinets are searched, piles of clothing and bedding are spread, and large boxes are searched. As in the primary search, report immediately any negative information, such as an unsafe condition.

Victim Removal

Once the victims are located, it will be necessary to remove them from the hazardous area or shelter them in place. More resources may need to be called as removal and treatment of victims will always pull personnel away from the firefight. An uninjured victim, or one with minor injuries, may only require assistance to walk to safety. One or two rescuers may be needed, depending on how much help is available and the size and condition of the victim. Once the victim is out of the IDLH, it is important to hand over the victim to EMS personnel and not abandon him or her without treatment **(Figure 5.5)**.

In fire conditions, smoke is the enemy to the airway of the victim. Remove the victim to cooler, fresh air as quickly as possible. Only move a victim through the hazard zone when there is no other choice.

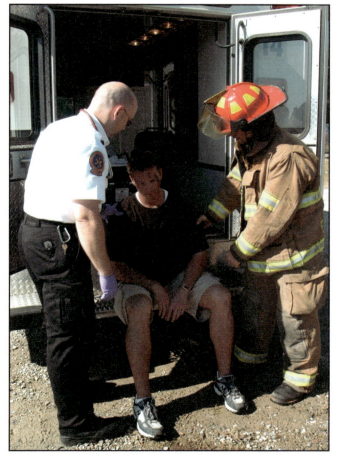

Figure 5.5 Once removed from the IDLH, victims must be transferred to EMS personnel for treatment or transport to a medical facility.

A danger of moving injured victims is that they may have severe orthopedic injuries. However, victims will not die from an orthopedic injury. They are more likely to die from the smoke and/or fire. In an extreme emergency, however, the possible spinal injury becomes secondary to the goal of preserving life. If it is necessary to perform an emergency move, the victim should be pulled in the direction of the long axis of the body — not sideways. Jackknifing the victim should also be avoided. If the victim is on the floor, pull on the victim's clothing in the neck or shoulder area. It may be easier to pull the victim onto a blanket and then drag the blanket. Being prepared to rescue a victim is a paramount job for firefighters. Personal webbing or rope should be carried and lifts and drags practiced. The first time you attempt to drag or move a person should not be an actual rescue in a fire.

> **NOTE:** The tactics and tasks performed on the fireground are essential to your safety and to the successful completion of the incident. Therefore, it is important that you and your crew learn and practice these tactics and tasks thoroughly. When the alarm sounds, it is too late to learn the correct procedures.

Figure 5.6 At a minimum, two rescuers are required to move an unconscious victim.

It is always better to have two or more rescuers when attempting to lift or carry an adult. One rescuer can safely carry a small child, but two, three, or even four rescuers may be needed to safely lift and carry a large adult. An unconscious victim is always more difficult to lift than a conscious one because the person is unable to assist in any way. With an unconscious victim, it is critically important that his or her head and neck be supported and stabilized to prevent further injury **(Figure 5.6)**.

Exposures

An exposure is any area to which fire could spread. Depending on the growth of a structure fire, it may have external or internal exposures. External exposures are other structures, equipment, vehicles, people, or natural features that would be endangered by the fire. Internal exposures are unburned or unaffected portions of the building interior that are threatened by fire growth. In some instances, protection may be a strategic goal. In others (particularly interior exposures), it may be a tactical objective supporting the strategic goal of containing the fire to the room/floor of origin. Protecting exposures involves activities that will prevent the spread of fire or products of combustion to them. Implementing the exposure tactic may also help meet the rescue and confinement objectives.

A thorough preincident survey will provide information on both interior and exterior exposures. Add to this information your knowledge of fire behavior, building construction, and information gathered on scene and you will enhance your ability to estimate the location and extent of the fire and project how it can spread.

Interior Exposures

Interior exposures to a fire are determined by the configuration of compartments and spaces and the location and arrangement of contents. Determining factors include:

- Type of construction (determined by building code) — Helps to determine how the fire and smoke will spread and the risk of collapse.

- Type of fire protection systems or lack of systems — Helps determine where fire can most easily be contained and how to support the system in controlling the fire.

- Floor plan (open, compartmentalized, attic, basement, etc) — Helps to determine how the fire and smoke will spread and if there are concealed spaces or barriers that will affect the spread **(Figure 5.7)**.

- Building occupancy type — Helps determine the amount and type of contents that will provide fuel for the fire and hazards for fire fighters.

- Number of stories — Indicates the possible extent of fire and smoke spread.

- Barriers to fire spread — Protection may include walls and doors that provide static barriers to fire spread and locations for hoseline placement.

- Heating, ventilation, and air conditioning systems — Can be potential avenues for smoke and fire spread. The HVAC should be controlled to minimize this potential risk. HVAC can sometimes be used to introduce fresh air to trapped victims.

- Weather conditions (wind velocity and direction) — Indicates direction, growth, and speed fire will spread.

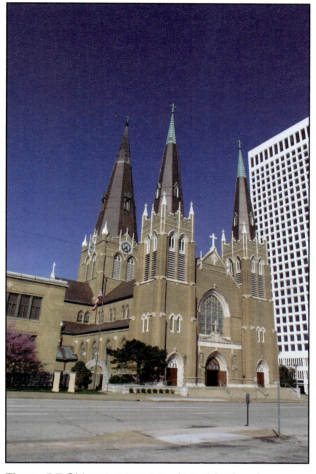

Figure 5.7 Older structures, such as this church, may have concealed spaces or voids created during expansion or renovation.

Exterior Exposures

Risk of fire extension to exposures adjacent to a fire building are determined by such factors as:

- Type of construction of the building on fire and exposed structure — Indicates how quickly the exposure will become involved in fire, the area where exposure protection is needed, where fire will most likely spread from and to, and how many BTUs will be given off.

- Type of fire protection systems or lack of systems in fire and exposed structure — Determines how much protection is available, how it can be controlled, and how it is supported.

- Building separation distances (often determined by building, fire, and sometimes zoning codes) — Indicates how rapidly the fire can extend from the fire building to the exposure **(Figure 5.8, p. 170)**.

Figure 5.8 In some communities, the space between structures may be so small that fire can easily move between buildings. *Courtesy of Bob Esposito.*

Figure 5.9 Not all exposures are buildings. Vehicles, vegetation, and propane storage tanks may also be endangered.

- Active and passive barriers to fire spread — Determines how long it will take for fire extension to the exposure.
- Nonstructural exposures — Indicates additional hazards such as flammable/combustible liquids storage, dry vegetation, vehicles, and Class A materials adjacent to the fire building **(Figure 5.9)**.
- Weather conditions (wind velocity and direction relative humidity and air temperature) — Indicates direction, growth, and speed that fire will spread, and which exposure is most in danger. Flying brands can cause fires several blocks away if not controlled.

Unless you can make a complete lap of the exterior of the entire structure, your perceptions of the fire situation may be in error. For instance, a fire reported in the early hours of the morning at a multistory warehouse in an older industrial district may create an orange-red glow and column of smoke from the back of the building. Does this mean that the fire is in the rear of the building, outside the building, or in an adjacent building across the alley? Even during daylight hours, you can mistake a column of smoke from a rubbish fire behind a structure for a fire in the rear of the structure. Or, you may not realize that the rubbish fire has spread into the building through a window or door. To the extent possible, you must verify your observations, based on fact, before committing your resources. You must also be ready to alter your perceptions after you get reports from interior and roof groups or divisions.

Your knowledge of the facts will guide you accurately in projecting how the fire will spread internally and externally and what exposures will be threatened by the fire. Some of the exposures, such as fire walls within the structure, will act as barriers to fire spread. Others, such as high stack piles of combustible contents, will add to the fuel load and increase the magnitude of the fire.

The tasks you assign to your resources to implement the exposure protection tactical objective will depend on the stage of the fire and whether the exposures are internal or external. The decision to assign exposure protection is indicated by many indicators as already discussed including weather, construction, fire size, crews, and water. The type of exposure protection will mostly be application of water, foam, or gel, and largely dependent on what is available and in what amount.

Exposure Tasks

Internal — Your initial task is to determine the location and extent of the fire. This can be accomplished through an investigation of the interior and a 360° walk around of the site when possible **(Figure 5.10)**.

Resources may be assigned to:
- Investigation
- Fire attack
- Water supply
- Evacuation
- Fire suppression systems (activation/support)
- Ventilation
- Removal/protection of exposed contents (salvage)
- Supporting the fire protection system in the fire building as needed

External — Tasks assigned where external exposures exist include:
- Perform a 360° walk around
- Evacuate fire building and exposed structures
- Establish water supply
- Activate master stream appliances as needed
- Relocate exposed equipment, vehicles, or materials (if possible)
- Support fire protection systems in fire building and exposures as needed

Figure 5.10 As part of the initial size-up, the IC performs a 360° survey to determine exposures.

Confinement

Depending on the location and extent of the fire, your attack mode may be offensive or defensive. Confinement includes those fire fighting operations required to prevent fire from extending from the area of involvement to uninvolved areas or structures. Confining the fire to the involved area may be used to assist with rescue by separating trapped victims from the fire.

The facts that you will need to consider when implementing confinement tactics are:

- Type of construction (determined by building code)
- Type of fire protection systems or lack of systems

- Building occupancy type
- Contents within a compartment
- Flammability and combustibility of contents
- Distance between involved and uninvolved contents
- Water supply vs. demand (fire flow)
- Number of stories
- Barriers to fire spread
- Heating, ventilation, and air conditioning systems
- Current location and stage of fire
- Resources immediately available
- Rescue or exposure tactics implemented
- Additional fuel or ignition sources
- Explosion potential (internal and external)
- Fire behavior

Figure 5.11 Initially, the only indication of a fire may be light smoke exiting the structure.

Size-up of the incident will give you an idea of how successful you will be in confining the fire to the compartment or structure of origin. Locating the origin of the fire and paths that it has taken or will take if it is allowed to expand may be difficult **(Figure 5.11)**. For instance, the appearance of heavy smoke or flames in a cockloft may indicate the direction of spread but not the source. Smoke showing from the roof may lead you to believe that the fire is there while in fact it is a basement fire. You should also be aware of excessive heat buildup in areas that do not have visible fire or smoke.

By knowing how the building is configured, the type of construction and contents, and the methods of heat transfer, you will be better able to determine how and where to stop the spread of fire. For example confinement can be accomplished by:

- Closing fire doors
- Using existing fire walls and fire protection systems
- Using drive-in or penetration nozzles from the roof or upper floor
- Shutting off the HVAC system
- Properly applying ventilation techniques

The proper application of ventilation techniques can help to draw out the heat and fire gasses from the structure, preventing the fire from spreading into other areas. In structures with continuous attics or cocklofts, **trench** cuts can be used as a defensive means of stopping fire spread. A trench cut is traditionally used on flat roofed buildings at a narrow point on the roof or adjacent to a fire wall on the side toward the fire. The roof is opened and the ceiling below is removed, either from above or below. In order to predict the spread of the fire, inspection holes (triangles or **kerf cuts**) must be made prior to the cutting of the trench on the fire side. Trench cuts can be made on pitched roofs adjacent to a fire wall by making the cut from the peak to a point at or near the eaves.

Trenching —In strip or trench ventilation, the process of opening a roof area the width of the building with an opening 2 foot (0.6 m) wide to channel out fire and heat.

Kerf Cut — the groove, cut, notch, or channel made by a saw or cutting tool.

As an Incident Commander, you must remember that trench cuts are time consuming and labor intensive. You must also realize that the trench cut is used after the primary ventilation hole is cut over the seat of the fire. The trench cut must be done with hoselines underneath and above, and access to the space through the ceiling.

NOTE: Trench cuts are of limited value and should not be attempted on large area roofs or roofs that are constructed of concrete or multiple layers of materials.

Confinement Tasks
- Access structure (forcible entry)
- Locate fire area
- Locate avenues of fire spread
- Ventilate over fire area, or apply horizontal ventilation to remove heat and products of combustion to the exterior of the building
- Place hoselines into use in appropriate attack strategy
- Establish water supply

Extinguishment

Extinguishing the fire is a natural extension of the confinement tactic. To accomplish this tactic, the extinguishing agent is placed on the seat of the fire or in the compartment containing the fire. If an offensive interior attack is required to extinguish the fire, ventilation may be used in conjunction as a means of cooling the compartment, removing heated fire gases, and improving visibility. Extinguishing may also be used to assist in rescue by removing the fire hazard altogether.

The facts that you will need to consider include:

- Fire behavior
- Building construction
- Length of time since the fire was discovered
- Fuel load including contents and structure
- Primary fuel type
- Location of the main body of fire
- Water supply
- Available resources and their equipment
- Attack mode in use
- Reflex time

Your perceptions may include:

- Effectiveness of fire confinement efforts
- Smoke color as you see it
- Smoke density

- Smoke level
- Water runoff
- Building stability indicators

Your size-up must provide the necessary facts to determine the best extinguishment method to apply. If the fire is contained in a small compartment surrounded by fire barrier walls, then a rapid application of water may produce a rapid extinguishment. However, if you are faced with a fire that is growing quickly with sufficient fuel and oxygen available, you may need large quantities of water supplied by large-caliber nozzles **(Figure 5.12)**.

In extinguishment as in rescue you must determine if the benefit outweighs the risk posed by advancing hoselines into the structure. If rescue is not necessary and the fire has expanded to consume much of the property, then using your resources to protect exposures and confine the fire to the barrier walls of the structure may be the least risky approach. Continual evaluation of the risk management plan is a must.

In addition to the tasks needed to confine the fire, extinguishment will require that your resources be committed to:

- Deploying master stream appliances, if applicable
- Advancing hoselines
- Selecting the most effective extinguishing agent
- Selecting the appropriate size and type of nozzle
- Selecting the appropriate type of ventilation method and deploying fans
- Locating the seat of the fire

In the majority of cases, water will be the primary type of extinguishing agent that you will have at your disposal. However, if you have a sufficient supply of Class A foam available for your initial attack, this agent may extinguish

Figure 5.12 A rapidly developing fire will require large quantities of water supplied by master stream appliances or large-caliber nozzles. *Courtesy of Bob Esposito.*

the fire rapidly with reduced water damage **(Figure 5.13)**. At the same time, the contents may require the use of Class B-type foam agents for complete and rapid extinguishment. Remember that additional water will add to weight stress to the structure increasing the collapse hazard.

Overhaul

Overhaul operations can begin once the main body of the fire has been extinguished and a safety assessment deems that it is safe to begin. Prior to beginning overhaul, an attempt to determine the cause and point of origin should be made. Overhaul operations can destroy or alter evidence of incendiary or accidental cause. Use caution to protect both evidence and personnel during this phase. According to Chief Layman, overhaul consists of:

Figure 5.13 When available, Class A or B foam extinguishing agents may be required to extinguish the fire.

- Searching for and extinguishing hidden or remaining fire including contents and concealed spaces
- Placing the building and its contents in a safe condition including the barricading of unsafe structures
- Determining the origin or cause of the fire
- Recognizing and preserving evidence of arson
- Identifying the area and point of origin

Some of the tasks performed to complete overhaul may be performed as part of extinguishment or even confinement. A traditional sign of professionalism in the fire service has been to completely extinguish all fire at an incident. There is nothing more unnecessary and embarrassing than to have to return to a scene because the fire has rekindled. The use of fresh or well-rested personnel or crews can minimize the possibility of this. A fire watch or drive-by checks of the fire scene may be needed.

It is also important from a liability standpoint to make the scene as safe as possible before returning the property to the owner. Legally, in some jurisdictions, the fire department takes possession of the site throughout fire suppression activities. Therefore, the site must be as safe as possible before the owner is allowed to enter or take possession. If the site is not safe, the structure should be barricaded to prevent unauthorized entry.

If it is suspected during overhaul that the fire was intentionally set, the structure must remain in the possession of the fire department or law enforcement agency. Once possession is returned to the owner, it may not be legally possible for the fire department to return to gather evidence, take photographs, or search for evidence or clues; the chain of custody has been broken. Fire officers who are certified to NFPA® 1021 should be able to determine the cause and origin of most fires **(Figure 5.14, p. 176)**.

NOTE: Your AHJ will determine who is responsible for determining fire cause and origin and when an investigator is required at the incident.

> **WARNING!**
> NO firefighter should ever be injured or killed during overhaul operations. ALWAYS ENSURE THAT THE STRUCTURE IS SAFE BEFORE OVERHAUL BEGINS. Overhaul is an extremely dangerous period on the fireground because:
> - Building structural members may have been weakened
> - Water weight has been added to the structure
> - High levels of carbon monoxide and hydrogen cyanide exist that can impair firefighters
> - Firefighters may be fatigued
> - Debris can create tripping hazards
> - Natural gas or electric lines may be present under debris or in enclosed spaces

Figure 5.14 In most jurisdictions, the company officer is responsible for determining the cause and origin of the fire. *Courtesy of Iowa State Training Bureau.*

Because overhaul generally occurs after the majority of the fire is extinguished, you may have more time to decide where and how the overhaul tactic will be performed. Because overhaul must be thorough, your decision must be based on solid facts. It is important to remember that when performing overhaul, you must be able to justify your actions to the owner.

The facts that you need to know before performing overhaul are:

- Building stability
- Fire behavior and extent of damage
- Building construction
  — Location of voids or concealed spaces
  — Construction materials used
  — Type of construction
- Estimated length of time the fire burned
- Indicators of fire ignition and spread (burn patterns)
- Indicators of intentional fire ignition
- Evidence of criminal activity
- Need for preserving evidence
- Type and combustibility of contents

Your senses may help you to determine where overhaul needs to be performed. You should:

- Look for signs of burning around openings such as electrical outlets or air-conditioning vents in walls, floors, and ceilings
- Look for smoke coming from the wall, ceiling, or floor cavities
- Feel wall surfaces with the back of the hand to determine if areas conceal fire
- Listen for the sound of burning materials in voids or concealed spaces
- Look for discoloration of walls in unburned compartments
- Look for temperature variations by using a thermal imaging camera (TIC)
- Look for smoke coming from ventilation systems

- Look for trailers or paths on flooring that may indicate the use of accelerants
- Look for deep-seated fires in contents such as bales of cotton or rolls of paper

However, never use your sense of smell to try to locate a fire or determine the presence of a combustible liquid. Breathing contaminated air can result in permanent injury or death. Respiratory protection must be worn during all overhaul operations unless the safety officer determines otherwise.

Given your knowledge of fire behavior and the dangers of fire spread in concealed spaces and the potential for rekindle, you should be able to project how a fire might spread unseen. The damage done by opening walls, floors, ceilings, and concealed spaces is far less than the damage from a rekindle of the fire. In addition, the rekindle would further endanger civilians working to clean up the area and firefighters who would be called to the new fire.

The risks associated with overhaul extend beyond the potential for a rekindle. Added risks include:

- Exposure to toxic or contaminated atmospheres
- Injury from sharp objects
- Energized electrical wiring
- Presence of gas lines
- Potential of structural collapse
- Injury or fatalities related to overexertion
- Injury or fatalities due to strokes or cardiac arrest
- Effects of heat stress
- Hazardous materials

These risks can be mitigated by taking the following steps:

- Wearing complete personal protective equipment (PPE) including appropriate respiratory protection unless or until the area has been declared free of noxious gases by a qualified person with appropriate equipment
- Ensuring the structural integrity of the building before committing resources to overhaul
- Providing rehabilitation facilities during the entire operation
- Providing on-site medical support
- Requesting additional personnel to replace initial assignment units, also known as a Fire Detail
- Securing utilities at all times
- Contracting with an outside company to use heavy equipment to remove debris (this activity may need to be coordinated with the owner)
- Shoring or securing structural members that may collapse
- Removing excess water that may contribute to collapse
- Using proper lifting, cutting, and prying techniques to reduce injuries
- Requesting assistance from utility companies

> **Warning!**
> Never remove respiratory protection or other items of personal protective equipment while performing overhaul operations. Respiratory protection of some form must be worn in contaminated atmospheres such as those created by a fire.

> **Overhaul Tasks**
>
> Property conservation during overhaul may include some of the following tasks or considerations:
>
> - Cutting holes in walls, ceilings, or floors
> - Moving or removing debris
> - Securing the scene to prevent access
> - Advancing hoselines and extinguishing hidden or remaining fires
> - Forcing entry into hidden or concealed spaces
> - Breaching walls
> - Reinforcing walls, ceilings, or floors
> - Shoring weak structural members
> - Removing excess water
> - Collecting or preserving evidence
> - Identifying and interviewing witnesses
> - Creating a sketch of the site
> - Making photographs
> - Requesting a fire investigator if warranted

Ventilation

Coordinated ventilation operations may be implemented at any time during the operation. This must be done in coordination with interior rescue or suppression activities. Ventilation is the systematic and coordinated channeling and removal of heated air, smoke, fire gases, or other airborne contaminants from a structure and replacing them with cooler and/or fresher air to reduce damage and to facilitate fire fighting operations. Early ventilation, if well-coordinated, will often prevent a flashover or backdraft.

Ventilation activities are traditionally associated with ladder/truck company duties, but this is not necessarily always the case. Many organizations may not have a ladder/truck company within their jurisdiction. If there is a delay or absence of a ladder/truck company, an engine company should be assigned to perform these functions. As the first-arriving officer, you may initiate ventilation to facilitate rescue or advance a hoseline to confine or extinguish a fire. Care must be taken when you ventilate a compartment or structure to prevent rapid fire expansion by the intake of fresh air. Hoselines and personnel must be in place when opening or making access points for ventilation.

Applying the proper type of ventilation at the correct time and place for maximum effectiveness will depend on your knowledge of fire behavior, your knowledge of the structure, and your ability to perform size-up. If ventilation is not implemented and coordinated correctly, the risk to firefighters, occupants, and the structure could be greatly increased **(Figures 5.15a-c)**.

The primary set of facts that you will have when deciding to ventilate a structure are:

- Fire behavior
  - Stage of fire

- Volume of smoke
- Movement of smoke
- Density of smoke
- Location of fire
- Potential fire spread
- Need for rescue, confinement, or extinguishment
- Available resources, including personnel and equipment
- Wind velocity and direction
- Building construction
  - Type of construction
  - Number of floors
  - Fire suppression system or lack of a system
  - Building layout
  - Building HVAC system

Figures 5.15a-c Ventilation may be accomplished through (a) horizontal, (b) vertical, or (c) positive pressure methods. (*b*) *Courtesy of Mathew Daly, (c) Courtesy of Iowa State Training Bureau.*

Your size-up will give you additional information based on your perception of the situation including:

- Smoke color
- Smoke density
- Smoke column
- Visible indicators of extreme fire development
- Sounds coming from the building
- Discoloration of window glass
- Condition of roof structure
- Estimated length of time fire has been burning

You must be able to estimate what will occur when ventilation is implemented and what will occur if ventilation does not happen. Introducing air into the superheated atmosphere can cause a flashover or backdraft, especially if introduced from lower floors or levels. Care must be taken to relieve the heat, pressure, and smoke in the structure without drawing in sufficient air to cause potentially extreme fire events.

Quite obviously, ventilation operations are risky in many ways. Guidelines for risk mitigation with ventilation that should be considered include the following:

- Personnel should not be on an unstable roof. They should work from aerial devices or ground ladders.
- Ventilation used at the wrong time or place can cause rapid fire development.
- Not ventilating in a timely fashion can cause the heat and smoke to increase within the compartment, which inceases the risk of flashover. This can also increase further fire and toxic gas extension, which results in greater risk to life and fire spread.
- Improper ventilation techniques can draw or force heat and smoke into unaffected portions of the structure.
- Improper use of positive pressure ventilation can cause rapid fire growth and flashover and can result in firefighter injury and death.

Regardless of whether you are combining ventilation with rescue, confinement, or extinguishment, the ventilation activities must be coordinated with the other activities. The two terms that are most associated with offensive ventilation are *vent for life* and *vent for fire*. Vent for life is an action that will improve the atmosphere by allowing smoke and heated gas to escape as well as allow fresh air to enter for potential victims. Use caution as this requires an aggressive and coordinated effort between suppression and rescue personnel. Vent for fire is a coordinated operation that is used to assist in fire extinguishment. Ventilation is performed at a point opposite the attack hoselines. Because any type of ventilation to an active fire can be dangerous, it is important that the IC tells the ventilation crew where to ventilate. It can also be used to assist in finding hidden, smoldering fires by allowing oxygen to enter as well as allowing smoke and heated gases to escape.

It is important to have hoselines in place in advance of this operation, as the oxygen allows the smoldering materials to flame. Although intended to help interior rescue efforts, if not used correctly ventilation can harm firefighters and occupants. Used cautiously, it is possible to use positive pressure ventilation (PPV) in an offensive strategy along with hoselines. However, this activity requires special training and a solid understanding of fire conditions and location. Inappropriate use of PPV can cause rapid fire spread and extreme danger to interior crews.

Ventilation Tasks

Tasks that may be required in order to implement the ventilation tactical objective include:

- Opening roof access panels, skylights, or vents
- Cutting vent holes
- Removing security bars on doors or windows
- Opening doors or windows opposite attack hoselines
- Placing hoselines in operation
- Placing PPV or smoke ejector fans in operation

Salvage

Although salvage operations are intended to meet the third strategic goal of property conservation, it is part of the initial operations and continues throughout the event. Salvage is defined as methods and procedures by which firefighters attempt to save property and reduce further damage from water, smoke, heat, and exposure. Salvage operations are performed during or immediately after a fire by removing property from a fire area, covering it, or using other means. Salvage tactical operations begin when resources are available to implement the tasks required to complete salvage.

Your preincident plan and size-up will provide the information needed to decide when to begin salvage operations and the areas that must be addressed first. While these decisions must be fact-based, you may also be confronted with the pressure of owners or occupants who believe more should be done to save their property. In commercial properties, merchandise, business records, computers, and electronic equipment may need to be removed rather than covered in place. At the same time, once the items are removed, they must be protected from the weather and theft.

In the case of residential properties, no preplan will exist. Therefore, your decision-making process should follow common sense. Items on lower floors should be protected from water damage or structural collapse including ceiling tiles, gypsum board, or light fixtures falling in. Items in adjacent compartments should be protected from smoke damage through ventilation or removal. Items in the fire compartment should be removed or covered if possible.

Some citizens are less fortunate than others and possibly those things that you may consider worthless may be a cherished treasure to others. In addition, many items found in residential properties and places of worship have emotional or sentimental value that exceeds their monetary value. Examples include religious artifacts and books, photographs and albums, children's artwork, and furniture. You will have to decide how much effort must be put into saving these types of items when you have limited resources available. Do not remove evidence of arson while performing salvage operations.

The preincident plan should provide an idea of the contents that will need to be removed or protected in place. In business and commercial properties, business records contained in filing cabinets, safes, and on computers are sometimes more valuable than a structure's merchandise or contents. The preincident plan should indicate if records are backed up at an off-site location.

Other facts you should be aware of include:

- Type and value of contents
- Location and type (hard copy or computer) of business records
- Type of fire suppression system (water-based or special-agent systems)
- Type of contents that are susceptible to water damage
- Location of fire or smoke barriers
- Height and floor space of structure
- Location of fire
- Ability to protect removed items outside the structure

When a preincident plan does not exist and witnesses are not around, you must use your own observations to decide what should be protected and in what manner. Look for such things as:

- Type and value of the structure
- Reports from owners or occupants
- Smoke conditions in structure
- Weather conditions (potential damage to items removed from structure)
- Difficulty involved in removing contents
- Difficulty involved with protecting removed contents

Figure 5.16 Water must be removed from structures to reduce water damage to property and to decrease the chance of structural collapse.

Determining the potential damage property may get from fire, heat, smoke, or water can be difficult. You must know how the fire is spreading, if confinement and extinguishment activities are successful, and where water will accumulate or travel. Structural collapse is a very big possibility when a large quantity of water has been applied to the fire. Removing the contents can require more resources than removing the water. You will need to know the most effective method for water removal that will take the least effort and cause the minimum of damage **(Figure 5.16)**. Salvage covers may not provide complete protection in the event of ceiling collapse or deep water.

Because salvage involves property, the risk assessment is based on the risk to firefighters compared to both the reduction in property loss and the affect on community morale. Saving a family heirloom is great for public relations, but it is not worth the risk to a firefighter who may be injured or killed in the effort.

182 Chapter 5 • Tactics

Salvage Tasks

Salvage tasks include:
- Use ventilation to remove heat, smoke, and fire gases
- Cause a minimum amount of damage during forcible entry
- Remove excess water to prevent damage or reduce risk of collapse
- Place waterproof covers over items that cannot be removed from the area
- Place waterproof floor runners to protect floor coverings
- Remove valuable items from the area
- Raise items above the water level in the area
- Shut off fire suppression systems when the fire is controlled
- Use limited water to effectively confine and extinguish the fire
- Use foam or gel extinguishing agents to efficiently confine and extinguish the fire
- Use non-water-based extinguishing agents to protect contents from water damage
- Remove debris from the area when the fire is extinguished
- Protect contents from vandalism or theft
- Cover any openings in roofs made for ventilation or caused by fire
- Secure all openings into the structure

Summary

To achieve the strategic priorities of life safety, incident stabilization, and property conservation, you must apply tactics. The tactical priorities created by Chief Lloyd Layman in the 1940s are still the most effective for structural fire fighting. They are encompassed in the term RECEO-VS which stands for Rescue, Exposure, Confinement, Extinguishment, Overhaul, Ventilation, and Salvage. You must decide on the tactic, or combination of tactics, and assign tasks to your crew and other units. Tactics such as ventilation may be used to support other tactics such as rescue or extinguishment. You must also be aware that changing conditions in the structure may require a change in tactics. Your primary concern, though, is life safety for yourself, your crew, your assigned units, and the public.

Review Questions

1. What are the two components of rescue?
2. What are the guidelines for searching safely?
3. What are the differences between primary and secondary searches?
4. What factors determine the risk of fire extension to both interior and exterior exposures?
5. What facts should be considered when implementing confinement tactics?
6. What factors should be considered regarding extinguishment tactics?

7. What is included in overhaul?
8. What facts must be known before performing overhaul?
9. How can risk related to ventilation be mitigated?
10. What tasks are included in salvage operations?

Residential Scenarios

Chapter Contents

Scenario 1 189	Scenario 6 229
Scenario 2 199	Scenario 7 237
Scenario 3 207	Scenario 8 243
Scenario 4 215	Scenario 9 249
Scenario 5 221	

Divider page photo courtesy of the Los Angeles Fire Department —ISTS.

chapter 6

Scenarios

This chapter contains nine hypothetical scenarios based on residential occupancy types that are found in most response areas. The scenarios contain best practices that can be applied to each type of situation. However, these are suggestions only and not hard and fast rules that must be applied in every incident.

Because these scenarios are examples to help you learn how to make decisions based on the information in the previous chapters, there are limitations placed on them. The resources available to you will arrive within 10 minutes. You are the Incident Commander (IC) and must allocate your resources to control/mitigate the incident to the best of your ability. The resources you have reflect those in your local department, including first-alarm assignments, standard operating procedures (SOPs), mutual aid, equipment, and staffing.

Residential Scenarios

Learning Objectives

1. Given information on a residential fire and resources available to your department, develop an incident action plan (IAP) for achieving the incident objectives.

Note to Instructors

The following scenarios contain recommended practices based on the facts provided and on information contained in the text of this manual. These scenarios are intended to be used as teaching aids. It is suggested that you develop scenarios specific to your department's resources, procedures, and response area. Your scenarios should contain structures, hazards, water supplies, photographs, and site plans that exist in your jurisdiction. Do not include probabilities, incident priorities, strategies, command options, or tactics. This information should be supplied by the students based on their understanding of the text. Your scenarios can be varied based on changes to the water supply (hydrants out of service, peak water demand times), delayed resources, weather conditions, minimum staffing, time of day, or local standard operating procedures (SOP).

Chapter 6
Residential Scenarios

Scenario 1

On Thursday, January 7, at 3:15 p.m., dispatch reports a residential structure fire at 411 S. Payne Drive. The dispatcher states that the owner reported the fire and that no occupants are in the structure. The owner stated that he was refinishing the woodwork in the structure when he noticed smoke and went outside to call 9-1-1. Current weather conditions are clear skies with a high-wind advisory.

On Arrival
You are the first-arriving officer. You see a one-story, single-family residence with heavy smoke and fire coming from the Delta (D) side of the structure.

Water Supply
One blue hydrant 1,500 gpm (5 680 L/min) is located directly across S. Payne Street

Weather
- Temperature — 48° F (9°C)
- Wind direction — North
- Wind velocity — 25 mph (40 km/mph)
- Humidity — 61%

360° View
Alpha Side

Bravo Side

Charlie Side

Delta Side

Chapter 6 • Residential Scenarios

Site Plan

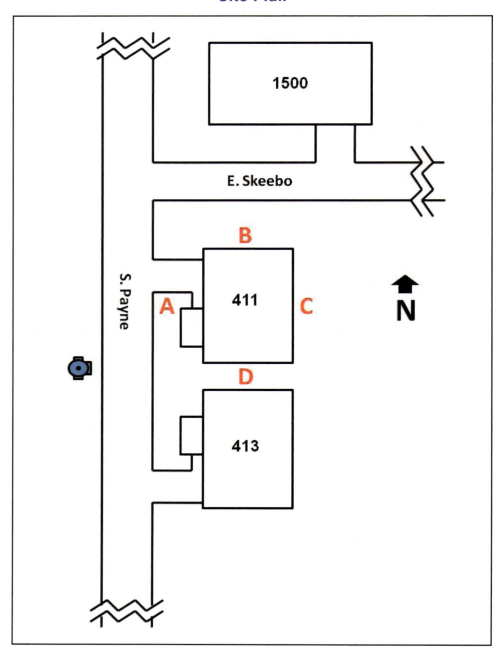

1. Size-Up

Based on the dispatch information and your visual observation of the scene on arrival, what are the facts and probabilities for this scenario?

Facts

- *Type V wood frame construction.*
- *Single-family residence that is unoccupied.*
- *Fire is visible in a front room area on the Alpha/Delta corner and a rear room on the Charlie/Delta corner.*
- *Smoke is coming from a window on Delta side, the gable vent, and from the vent openings in the roof decking on Charlie side.*

- *Wind is strong out of the north.*
- *Water supply is close to structure with one 1,500 gpm (5 680 L/min) available.*
- *There is an exposure on Delta side.*
- *The house is built on a foundation with vents visible. This means there is a wood floor deck with a crawl space underneath.*
- *There is a gas service meter at the Alpha/Delta corner of the house. It is close to the front room which is involved in fire.*

Probabilities
- *House appears to be new construction with probable lightweight construction features. A lack of landscaping increases the probability it is new construction.*
- *Reporting party, the owner, stated that he was doing woodworking in the house at the time of the fire. Hazardous materials may be present and fire growth and extension may be accelerated due to the wood finishes being used.*
- *There is radiant heat transfer to the exposure on Delta side.*
- *Fire has probably extended to the attic area as heavy black smoke is coming from the gable vent, Alpha side eaves, and from vent openings in the roof decking on Charlie side. There is some opening in the ceiling which has allowed the fire to extend.*
- *Lightweight trusses in the attic are being exposed to heat and flames due to extension.*
- *Fire has extended to multiple rooms (compartments) since smoke is coming from a window on Delta side. Flames are visible in that same room on Charlie side, and flames can also be seen in front-room windows on Alpha and Delta side.*
- *There is utility service in the house including electricity and gas.*

2. Incident Priorities

What are your considerations for addressing the incident priorities of life safety, incident stabilization, and property conservation?

Life Safety
- *The owner has reported that no one is in the structure. Also, the house appears to be unoccupied and still under construction. The probability of anyone requiring immediate rescue is low.*

Incident Stabilization
- *Two immediate situations must be addressed in incident stabilization: the fire in the structure and protecting the exposure.*
- *There is significant fire extension in the house involving multiple rooms and the attic. By extinguishing this fire, the threat to the exposure is also addressed.*

Property Conservation
- *From the information provided and what can be observed, it is unlikely the house is occupied. Consequently, there should be little property, such as home furnishings, at risk from the fire. There may be construction tools and equipment inside which are at risk.*

- *The exposure must also be considered at risk from the fire, specifically the radiant heat and brands coming from the attic and Delta side window. Extinguishing the fire eliminates this risk.*

3. Operational Strategy

What are your considerations when selecting your operational strategy?

- *Your strategy choice may depend on how many resources you will have quickly available at the scene. There is a direct route into the fire area through the front door which would make an offensive operation straightforward. However, there is a significant amount of fire involvement in at least one other room and most likely the attic. If an offensive strategy is selected, adequate resources must be available to quickly confine and extinguish the fire in both rooms as well as addressing any fire extension to the attic.*

- *Another consideration is the exposure on the Delta side. If adequate resources are not available to quickly confine and extinguish the fire using an offensive attack, consideration must be given to protecting the Delta side exposure through a defensive attack or focusing resources on protecting the exposure.*

- *A water supply is close to the structure that should be adequate to support an offensive attack or a defensive strategy for exposure protection.*

4. Command Options

What are your considerations when selecting a Command option: Investigation, Fast Attack, or Command?

- *Several conditions must be considered when selecting the Command option. The first and most important is the stability of the roof; that is, the potential for collapse due to the wind-driven fire extension and lightweight construction. The second is the potential for fire spread to the exposure. The Command option selected must ensure that these conditions are being monitored visually from outside the structure.*

5. Tactics

What are your considerations, if any, for the tactics (RECEO-VS) used to accomplish your incident priorities?

Rescue

- *The integrity of the structure must be considered when conducting a primary and secondary search. There may be a collapse hazard from the roof or ceiling finish due to fire extension and lightweight construction. Also, there is a crawl space underneath the house and the integrity of the floor should be monitored by the crews conducting the searches.*

- *Due to the risk of fire exposure to the house on Delta Side, consideration should be given to doing a primary search of the exposure or at least alerting the residents of the situation and having them leave their house until the situation is stabilized.*

Exposure

- *The exposure on Delta Side can be addressed in two ways: by extinguishing the fire in the structure or by placing a line on the Delta side of the exposure.*

- *The wind direction and speed are intensifying the fire and increasing the amount of heat reaching the exposure. This will shorten the time required for fire to ignite in the exposure and extension to occur. Because of the wind conditions, if the exposure does ignite, rapid fire extension should be expected.*
- *If a handline is used to protect the exposure, the amount of water available for attacking the fire will be reduced. Depending on resources and the amount of lines used to attack the fire, an additional hydrant or other water source may be required.*
- *The gas service meter at the Alpha/Delta corner of the house should be considered an exposure, at least from heat exposure from the fire in the living room. If there is fire impingement on the meter, consideration should be given to cooling the meter with short bursts of water. The local utility company should be contacted immediately.*

Confinement
- *The wind will cause the fire to extend quickly on the leeward side of the structure, that is, towards the Delta side. The wind may reduce the rate of extension to the windward side, that is, the Bravo side.*
- *Consideration must be given to confinement in at least three areas or compartments of the structure: the living room where the fire is visible, the back room on the Charlie/Delta corner, and the attic. If fire has extended into the crawl space, it must also be considered a compartment in which the fire must be confined.*
- *If the choice is made for an exterior attack due to a lack of available resources, consideration must be given to potential extension of fire into the unburned areas.*
- *It appears that entrance can be made through the front or back doors of the structure.*

Extinguishment
- *If the initial attack is made through the front door, consideration must be given to the integrity of the front porch. There is visual indication that fire may have extended into that concealed space. It should be monitored by the Incident Commander or a crew member.*
- *Any crews working inside, especially for initial confinement and extinguishment, must maintain situational awareness of the potential for roof or ceiling finish collapse and an unstable floor deck due to fire damage.*
- *Consideration must be given to quickly accessing the attic area during extinguishment to check for fire extension and, if necessary, to quickly extinguish any fire. Care must be taken to avoid or limit crews advancing hoselines into areas on the ground floor while there is active fire in the attic above them.*
- *If resources are limited and an interior attack is not possible, consider attacking the fire from the exterior. If an exterior attack is chosen, it is critical that other crews are not inside the structure working or in a location where they could be injured by the fire stream or fire and heat. Also, care must be taken to prevent the fire from being driven by the fire stream into unburned areas of the structure. Finally, it may not be possible to safely and effectively extinguish the fire in the attic from the exterior due to the wind conditions and the location of the fire.*

Overhaul
- *During overhaul, consideration must be given to the structural integrity. Fire may have damaged the lightweight trusses and the floor deck and/or joists. It may be necessary to limit activities in any areas where there is a potential collapse. Also, the presence of high levels of toxin (for example, carbon monoxide and hydrogen cyanide) in the atmosphere make the use of SCBA mandatory until the Safety Officer determines the risk has been eliminated.*

Ventilation
- *If horizontal ventilation is conducted, especially by opening the living room window on Delta side, consideration must be given to potential impact to the exposure.*
- *The velocity and direction of the wind must be considered. The wind will increase the amount of air moving through the structure and aid in ventilation. However, the same condition will greatly increase the growth and extension of the fire, especially in the attic.*
- *For vertical ventilation, consideration must be given to the integrity of the roof deck, especially in the area above the fire. The color and quantity of smoke coming from the gable on Delta side strongly indicates the fire has already extended to the attic. This means that at least some of the trusses are involved and weakened. Any firefighter who may be on the roof in that area is at risk of falling through the roof into the attic.*

Salvage
- *There is probably no unusual consideration for salvage. If construction tools and equipment are inside they should be treated with care as they may be expensive to replace and susceptible to water damage.*

6. Unusual Hazards or Conditions

What are any unusual hazards or conditions which must be addressed? How will you address those hazards and/or conditions?

- *Unknown quantity and type of hazardous materials. The reporting party stated that he was refinishing woodwork in the house. This probably involved some type of flammable finish, especially since a fire occurred during this process. There may be multiple gallons of flammable or combustible finishers, cleaners, etc. in the house.*
- *Exposure of gas service meter to heat and flame. The gas service meter at the Alpha/Delta corner must be protected by some method if it becomes exposed to heat or flame.*
- *Lightweight construction. This is a hazard which must be considered due to the age of the structure. Lightweight members may have been used in both the roof and floor construction. The structural integrity must be monitored by the Incident Commander and all firefighters working in or around the structure. Also, fire attack and rescue crews must be alert for any openings or weak areas in the floor caused by fire damage.*
- *Wind conditions. The wind velocity and direction will quicken the spread of the fire and the amount of resulting damage to structural content and members. It also increases the risk of fire extension to the Delta side exposure.*

- *Porch over front door. This is only a hazard based on the visual indicators of smoke coming from the eaves of the porch. This is typically a concealed space which makes extinguishment difficult. Also, as fire weakens the structure, the porch could collapse on any firefighters on the porch or obstruct access through the front door. It should be monitored by the Incident Commander and any firefighters working around that area.*

Notes

Notes

Scenario 2

On Friday, December 22, at 8:15 p.m., dispatch reports a structure fire at a residence located at 1105 E. McElroy Court. The Dispatcher states that a neighbor reported the fire and is not sure if the occupants are home. The current weather condition is light snow flurries following a heavy snowfall during the day. Roads have been reported slick with snow accumulations of 2 to 4 inches (50 to 100 mm).

On Arrival
You are the first-arriving officer. You see a one-story, single-family residence with heavy smoke and fire coming from the open garage door on the Alpha side of the structure adjacent to the corner with the Bravo side.

Water Supply
One blue 1,500 gpm (5 680 L/min) hydrant is located at the corner of 11th Street and E McElroy Court.

One blue 1,500 gpm (5 680 L/min) hydrant is located at the corner of Oklahoma Street and E McElroy.

Weather
- Temperature — 29°F (-2°C)
- Wind direction — Northwest
- Wind velocity — 15 mph (24 km/h)
- Humidity — 100%

360° View
Alpha Side

Bravo Side

Charlie/Delta Sides

Site Plan

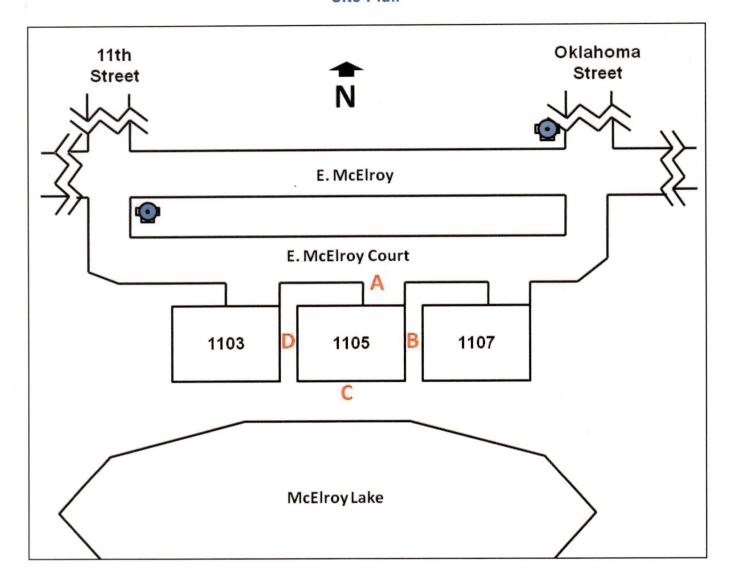

1. Size-Up

Based on the dispatch information and your visual observation of the scene on arrival, what are the facts and probabilities for this scenario?

Facts

- Type V wood frame construction.
- The house appears to be built on a slab foundation.
- Single-family residence that may be occupied.
- Fire is visible in the garage on the Alpha side of the structure.
- Smoke is coming from the garage door but is not apparent in the living area of the structure.
- Wind is strong out of the northwest blowing into the open garage door. This can create a wind-driven fire if interior door is opened causing the fire to extend into the rest of the structure.

- *Water supply is close to structure with two 1,500 gpm (5 680 L/min) hydrants available.*
- *There are exposures on Bravo and Delta sides.*
- *Smoke buildup on roof and around structure.*

Probabilities

- *House appears to be new construction with probable lightweight construction features. A lack of landscaping increases the probability it is new construction.*
- *Reporting party, a neighbor, did not confirm if the residents are home. The time of day, weather conditions, combined with the open garage door makes it likely that someone is home.*
- *There is possible radiant heat transfer to the exposure on Bravo side.*
- *Lightweight trusses in the attic above the garage may be exposed to heat and flames due to extension.*
- *Intensity of fire may have affected the garage overhead door creating a likely collapse hazard.*
- *Fire seems to be confined to the garage area. The wall separating the garage and the living area should provide a barrier to fire extension.*
- *The connecting door between the garage and the house may be open or closed. If it is open, it will provide an avenue for fire extension. If it is closed, it will help confine the fire to the garage.*
- *There is utility service in the house including electricity.*

2. Incident Priorities

What are your considerations for addressing the incident priorities of life safety, incident stabilization, and property conservation?

Life Safety

- *Without confirmation that the structure is unoccupied, it must be assumed that someone is in the living area. Search activities can be combined with fire attack or assigned to later arriving units.*

Incident Stabilization

- *Two immediate situations must be addressed in incident stabilization: the fire in the structure and protecting the exposure.*
- *Rapid fire attack and extinguishment can accomplish all three incident priorities. Placement of the attack lines will depend on whether the connecting door is open or closed. If it is open, the lines can be advanced through the living area and fire streams directed from the interior, forcing the fire out the garage door. If the door is closed, the lines can be directed through the garage door.*

Property Conservation

- *The primary concern is to prevent the fire from extending into the attic and living areas. Rapid extinguishment can prevent both and limit damage to the contents of the garage.*

- *The exposure on the Bravo side is a consideration and rapid fire extinguishment eliminates this risk.*

3. Operational Strategy

What are your considerations when selecting your operational strategy?

- *Your strategy choice may depend on how many resources you will have quickly available at the scene. There is a direct route into the fire area through the garage door, which would make an offensive operation straightforward. If an offensive strategy is selected, adequate resources must be available to quickly confine and extinguish the fire.*

- *A water supply is close to the structure that should be adequate to support an offensive attack or a defensive strategy for exposure protection.*

4. Command Options

What are your considerations when selecting a Command option: Investigation, Fast Attack, or Command?

Several conditions must be considered when selecting the Fast-Attack Command option. The first and most important is the stability of the retracted overhead door and ceiling in the garage and the lightweight construction of the roof trusses. The second is the potential for fire spread to the interior exposure caused by an open door, the wind direction, and fire streams directed into the garage. If an interior attack from the connecting door is made, these conditions must be monitored visually from outside the structure.

5. Tactics

What are your considerations, if any, for the tactics (RECEO-VS) used to accomplish your incident priorities?

Rescue

- *The integrity of the structure must be considered when conducting a primary and secondary search. There may be a collapse hazard from the ceiling finish and retracted overhead door due to direct flame contact on the door supports and lightweight construction of the roof trusses. Search and rescue of the living area should not present a risk to responders.*

Exposure

- *The primary exposure hazard is the living area inside the structure. Placing an attack line in the connecting door and directing the fire stream from that location can protect this area.*

- *The exposure on Bravo side can be addressed in two ways: by extinguishing the fire in the structure or by placing a line on the Bravo side of the exposure.*

- *The wind direction and speed can drive the fire back into the garage rather than at the structure on the Bravo side.*

Confinement

- *Structural components including the fire separation walls and ceiling materials can help to confine the fire to the garage. Placement of an attack line in the connecting door from the living area can limit or prevent extension into the living area.*

- *If the choice is made for an exterior attack from the garage door due to a lack of available resources, consideration must be given to potential extension of fire into the attic and living areas.*
- *It appears that entrance can be made through the front or back doors of the structure.*

Extinguishment

- *If the initial attack is made through the garage door, consideration must be given to the structural integrity of the retracted overhead door supports. The door must be considered a collapse hazard throughout the operation. Even if the door supports are intact, the door must be blocked open to prevent it from closing on a charged hoseline and interrupting the flow of water at a critical time.*
- *Any crews working inside, especially for initial confinement and extinguishment, must maintain situational awareness of the potential for roof or ceiling finish collapse in addition to the overhead door.*
- *Consideration must be given to quickly accessing the attic area during extinguishment to check for fire extension and, if necessary, to quickly extinguish any fire. Care must be taken to avoid or limit crews advancing hoselines into areas on the ground floor while there is active fire in the attic above them.*
- *If resources are limited and an interior attack is not possible, consider attacking the fire from the exterior. If an exterior attack is chosen, it is critical that other crews are not inside the structure working or in a location where they could be injured by the fire stream or fire and heat. Also, care must be taken to prevent the fire from being driven by the fire stream into unburned areas of the structure. Finally, it may not be possible to safely and effectively extinguish the fire in the attic from the exterior due to the wind conditions and the location of the fire.*

Overhaul

- *During overhaul, consideration must be given to the structural integrity. Fire may have damaged the retracted overhead door supports and lightweight trusses and/or joists. It may be necessary to limit activities in any areas where there is a potential collapse.*

Ventilation

- *If horizontal ventilation is conducted, especially by opening the living room window on Delta side, consideration must be given to potential impact to the exposure and extension of fire.*
- *The velocity and direction of the wind must be considered. The wind will increase the amount of air moving through the structure and aid in ventilation. However, the same condition will greatly increase the growth and extension of the fire, especially in the attic.*
- *For vertical ventilation, consideration must be given to the integrity of the roof deck, especially in the area above the fire. Snow and ice buildup on the roof will create an additional hazard for personnel working on the roof.*

Salvage

- *There is probably no unusual consideration for salvage. Contents of the garage may be removed to prevent further damage. The overhead door must be removed or supported to prevent collapse.*

- *As part of the salvage activities, it may be necessary to shut off some or all of the utilities. If shutting off the utilities is necessary, consider the effect of shutting off the source of heating for the house.*

6. Unusual Hazards or Conditions

What are any unusual hazards or conditions which must be addressed? How will you address those hazards and/or conditions?

- *Unknown quantity and type of hazardous materials. There may be multiple gallons of flammable or combustible finishers, cleaners, gasoline, or propane tanks stored in the garage.*

- *Lightweight construction. This is a hazard which must be considered due to the age of the structure. Lightweight members may have been used in both the roof and floor construction. The structural integrity must be monitored by the Incident Commander and all firefighters working in or around the structure.*

- *Weather conditions and the affect on personnel. The below freezing temperature plus the wind chill factor will reduce the physical strength of personnel during a long incident. Frostbite will also be a threat. Rehabilitation and medical assistance as well as relief personnel will be required.*

Scenario 3

On Tuesday, January 3, at 7:45 a.m., dispatch reports a residential structure fire at 110 Mountain Ranches Road. The dispatcher states that a passerby reported the incident and was not sure if occupants were home. The current weather condition is snowy.

On Arrival
You are the first-arriving officer. You see a two-story, single-family residence with heavy smoke coming from the basement window and soffit of the structure. Neighbors report that they saw family members at a second-story window, and they did not see anyone exit the structure.

Water Supply
One orange 800 gpm (3 780 L/min) hydrant is located at the corner of Sooner Road and W. McElroy Street.

Weather
- Temperature — 15° F (-10°C)
- Wind direction — Northeast
- Wind velocity — 4 mph (6 km/h)
- Humidity — 15%

360° View

Alpha Side

Bravo Side

Charlie Side

Delta Side

Site Plan

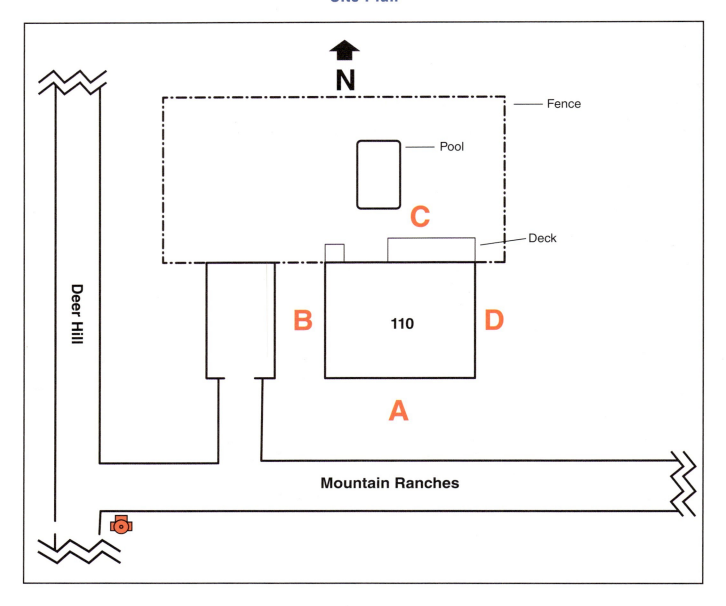

1. Size-Up

Based on the dispatch information and your visual observation of the scene on arrival, what are the facts and probabilities for this scenario?

Facts

- *Type V wood frame construction.*
- *The structure is multistory including a basement and two floors aboveground.*
- *Single-family residence that is reported to be occupied.*
- *Heavy smoke coming from the basement window and soffit of the structure.*
- *Water supply is close to structure with one 800 gpm (3 028 L/min) hydrant available.*
- *There is an exposure on the Bravo/Alpha sides.*

- *Heavy snow buildup on the ground and on a covered swimming pool on the Charlie side.*

Probabilities
- *There could be radiant heat transfer to the exposure on Bravo side.*
- *Primary fire located in basement.*
- *Fire may be extending up wall cavities and into attic.*
- *Witness reports and time of day indicate that occupants may be trapped on the second floor.*
- *Basement fire may have weakened exposed floor joists on fire floor above fire location creating a collapse hazard for part of the floor.*
- *There is utility service in the house including electricity and propane gas.*

2. Incident Priorities
What are your considerations for addressing the incident priorities of life safety, incident stabilization, and property conservation?

Life Safety
- *Neighbors reported seeing residents in second floor windows.*
- *Search and rescue are the primary priorities. A charged hoseline must be placed inside the structure to protect search teams from fire extension in the basement and into the attic.*

Incident Stabilization
- *The primary access to the fire is the interior stairwell. The initial attack line would advance through the front door and to the basement door. Floor must be checked for stability throughout advancement of hose.*
- *The use of a piercing or drive-in nozzle should be considered if the floor is too weak to support an attack line and crew.*
- *If an interior attack is not possible due to structural instability, attack lines can be directed through the basement windows.*

Property Conservation
- *Rapid fire extinguishment in the basement is the most effective way to protect the contents of the house above the fire. Some items may need to be removed, but the risk to firefighters of floor collapse must be considered.*
- *Ventilation can be used to decrease smoke contamination and damage.*
- *The exposure must also be considered at risk from the fire, specifically the radiant heat and brands coming from the Bravo casement window. Extinguishing the fire eliminates this risk.*

3. Operational Strategy
What are your considerations when selecting your operational strategy?

- *Your strategy choice may depend on how many resources you will have quickly available at the scene. There are direct routes to the interior basement door through the front and back doors making an offensive operation straightfor-*

ward. *If an offensive strategy is selected, adequate resources must be available to quickly confine and extinguish the fire while performing search and rescue of the upper floors.*

- *Another consideration is the exposure on the Bravo side. Rapid confinement and extinguishment is the most effective way to protect the exposure.*

- *A water supply is close to the structure that should be adequate to support an offensive attack or a defensive strategy for exposure protection.*

4. Command Options

What are your considerations when selecting a Command option: Investigation, Fast Attack, or Command?

- *Several conditions must be considered when selecting the Command option. The first and most important is the life safety priority. Resources must be allocated for search and rescue as well as fire extinguishment. The Command option selected must ensure that these conditions are being monitored visually from outside the structure.*

5. Tactics

What are your considerations, if any, for the tactics (RECEO-VS) used to accomplish your incident priorities?

Rescue

- *The integrity of the floor must be considered when conducting a primary and secondary search. The hazard of floor collapse is high due to the visual evidence of fire or, at least, smoke extension to the attic level.*

- *The possibility is high that the residents seen on the second floor have been overcome by smoke or are disoriented.*

- *Access to the second-floor windows by way of ground ladders is also possible if the necessary resources are available.*

- *The number of resources committed to search and rescue will exceed those assigned to fire extinguishment.*

Exposure

- *Interior exposures can be protected by the primary attack hoseline through the front or back door.*

- *The garage on the Bravo side can be protected by a hoseline if the fire continues to spread out the window and up the exterior wall.*

- *The propane gas cylinder on the Charlie side of the house should be considered an exposure but only if the fire spreads through the structure.*

Confinement

- *Rapid placement of a hoseline onto the first floor can confine the fire to the basement. Depending on the stability of the floor, the hose can be used to attack the fire down the stairs or to supply a piercing or drive-in nozzle through the floor.*

- *Opening the inside of the exterior walls on Bravo side can help in preventing fire extension to upper floors.*

Extinguishment

- *If the initial attack is made through the front door, consideration must be given to the structural condition of the first floor. If the floor is too weak to support firefighters, then consider using piercing or drive-in nozzles or an exterior attack through the basement windows.*
- *Any crews working inside, especially for initial confinement and extinguishment, must maintain situational awareness of the potential for floor collapse due to fire damage.*
- *If resources are limited and an interior attack is not possible, consider attacking the fire from the exterior. If an exterior attack is chosen, it is critical that other crews are not inside the structure working or in a location where they could be injured by the fire stream or fire and heat. Also, care must be taken to prevent the fire from being driven by the fire stream into unburned areas of the structure.*

Overhaul

- *During overhaul, consideration must be given to the structural integrity. Fire may have damaged the lightweight trusses in the attic and the floor joists over the basement. It may be necessary to limit activities in any areas where there is a potential collapse.*
- *The Bravo side wall as well as any walls directly over the fire area must be opened on the first and, if indicated as needed, on the second floor. Interior access to the attic must be made to access any fire extension in that area.*

Ventilation

- *Opening windows on the first floor and using natural ventilation to remove smoke from the immediate area will improve visibility for search and attack teams.*
- *When the fire is extinguished, an ejector can be placed in the basement windows to pull smoke from that area.*
- *Use forced ventilation at the front door to provide fresh air once the fire is extinguished.*

Salvage

- *Properly applied ventilation techniques can reduce the potential for smoke damage.*
- *Water should be removed from the basement when the fire is extinguished.*
- *Salvage covers and floor runners can be used during the operation, if the floor is stable, to protect contents on first and second floors.*

6. Unusual Hazards or Conditions

What are any unusual hazards or conditions which must be addressed? How will you address those hazards and/or conditions?

- *Potential floor collapse from basement fire.*
- *Possible fuel leak from propane piping to furnace in basement.*
- *Personnel must work from above the fire.*

- *Slippery conditions on roof, creating hazards during vertical ventilation.*
- *Covered pool on Charlie side close to house, creating fall hazard.*
- *Snow and ice conditions can slow response times and delay hoseline deployment.*
- *Unknown contents stored in basement.*

Notes

Scenario 4

On Wednesday, August 15, at 2:00 p.m., dispatch reports a mutual-aid request for a residential structure fire at 2211 East Country Trail Road. This residence is located just outside your department's jurisdiction. The current weather condition is clear with high wind advisory.

On Arrival
You are the first-arriving officer. You see a manufactured single-family residential trailer with fire and heavy smoke coming from Charlie side of the structure.

Water Supply
No hydrant water supply is available.

Weather
- Temperature — 90° F (32° C)
- Wind direction — North
- Wind velocity — 20 mph (32 km/h)
- Humidity — 48%

360° View

Alpha/Delta Side

Bravo/Charlie Side

Site Plan

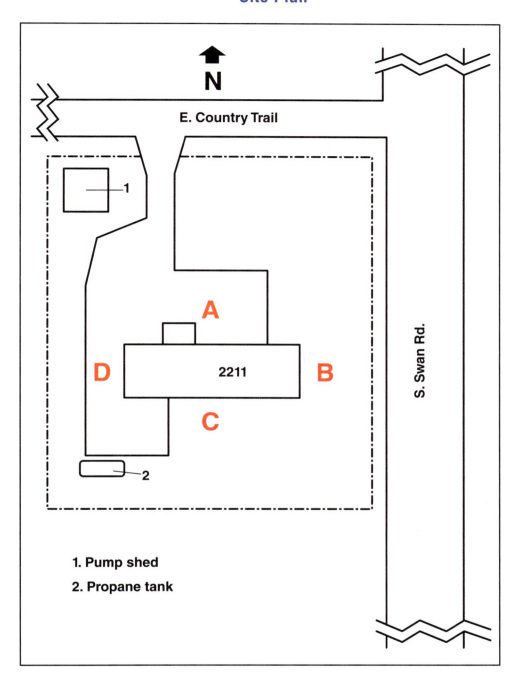

1. Pump shed
2. Propane tank

1. Size-Up

Based on the dispatch information and your visual observation of the scene on arrival, what are the facts and probabilities for this scenario?

Facts

- *Manufactured home, approximately 8 x 40 feet (2.4 m by 12.2 m), Type V wood frame, metal clad construction.*
- *Single-family residence that appears to be unoccupied.*
- *Wind is strong out of the north.*

Chapter 6 • Residential Scenarios **217**

- *There is no water supply available.*
- *Dry grass and ground cover surround the structure.*
- *There is a propane tank approximately 25 feet (7.5 m) from the Charlie side.*
- *The house is built on a raised foundation meaning that there is a wood floor deck with a crawl space underneath.*

Probabilities

- *Condition of structure may indicate pre-standard construction.*
- *Lack of occupants or vehicles may indicate structure is unoccupied or occupants are not savable.*
- *Wind velocity and dry vegetation may increase exposure hazard for wildland fire.*
- *Lack of water supply may require implementation of water shuttle or request for water tender.*

2. Incident Priorities

What are your considerations for addressing the incident priorities of life safety, incident stabilization, and property conservation?

Life Safety

- *The structure is unoccupied, based on visual evidence. Also, the environment appears to be untenable and the likelihood of a trapped resident being alive is low.*

Incident Stabilization

- *The amount of fire visible and the lack of immediate water supply would indicate the need to confine the fire to the structure and prevent it from extending to the surrounding grasslands.*

Property Conservation

- *The amount of fire visible would indicate that the structure and contents are beyond saving.*

3. Operational Strategy

What are your considerations when selecting your operational strategy?
- *Due to the advanced stage of the fire, lack of water supply, and weather conditions, a defensive strategy is the most appropriate.*

4. Command Options

What are your considerations when selecting a Command option: Investigation, Fast Attack, or Command?

- *Because a defensive strategy is being implemented, the appropriate Command option in this scenario is "Command."*
- *The IC must be in a position to direct all resources in extinguishment and exposure protection.*

5. Tactics

What are your considerations, if any, for the tactics (RECEO-VS) used to accomplish your incident priorities?

Rescue

- Search must wait until fire extinguishment is complete due to the advanced stage of the fire.

Exposure

- The primary exposure is the dry grass surrounding the structure on all sides.
- Resources must be assigned to spark patrol downwind of the fire as well as areas adjacent to the fire.
- The propane tank can be protected with the same resources that are protecting the adjacent areas.

Confinement

- The fire is confined to the structure itself through exposure protection and extinguishment of the fire.

Extinguishment

- The advanced stage of the fire will require the application of large quantities of water while using an exterior attack.
- Care must be taken to maintain an adequate supply of water for exposure protection while still providing enough to extinguish the fire.

Overhaul

- Overhaul will only be implemented when the fire is extinguished to ensure that debris will not rekindle.

Ventilation

- Ventilation is not required in this scenario.

Salvage

- Salvage is not required in this scenario.

6. Unusual Hazards or Conditions

What are any unusual hazards or conditions which must be addressed? How will you address those hazards and/or conditions?

- The high winds and heat are a factor for fire spread to the surrounding areas.
- The heat and humidity increase the risk to firefighters for heat stress.
- Lack of immediate uninterrupted water supply will increase the time required to terminate the incident.

Notes

Scenario 5

On Monday, October 4, at 4:30 p.m., dispatch reports a residential structure fire at 834 E. Skeebo Road. The dispatcher states that the call came from the owner who is at work saying her daughter called her to report the fire and is trapped in her room on the second floor.

On Arrival
You are the first-arriving officer. You see a split-level, single-family residence with heavy smoke and fire coming from the Alpha side of the structure.

Water Supply
One blue 1,500 gpm (5 680 L/min) hydrant is located one block north at the corner of E. Kelly Street and S. Payne.

One blue 1,500 gpm (5 680 L/min) hydrant is located ½ block South on S. Payne Street.

Weather
- Temperature – 54°F (12°C)
- Wind direction – Southwest
- Wind velocity – 5 mph (8 km/h)
- Humidity – 75%

360° View

Alpha Side

Bravo Side

222 Chapter 6 • Residential Scenarios

Charlie Side

Delta Side

Chapter 6 • Residential Scenarios

Site Plan

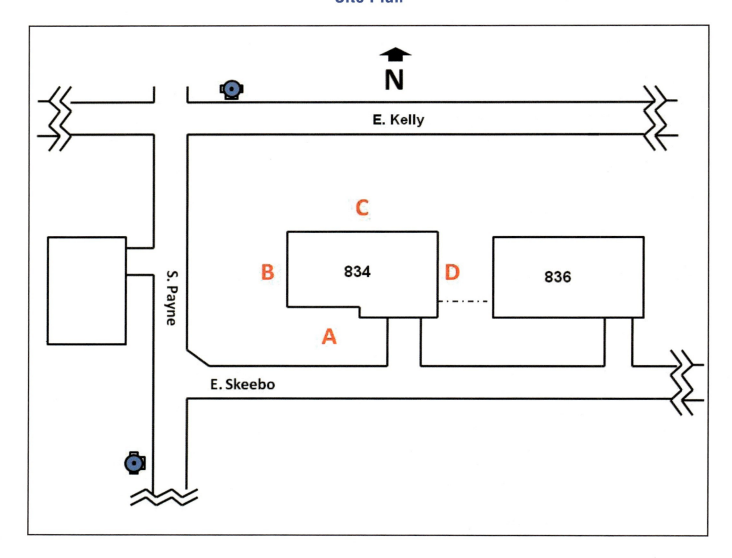

1. Size-Up

Based on the dispatch information and your visual observation of the scene on arrival, what are the facts and probabilities for this scenario?

Facts
- *Type V wood frame construction.*
- *The design of the structure is split level with a living area above the garage.*
- *Single-family residence that is occupied.*
- *Fire is visible from a window on the Alpha side of the structure.*
- *Smoke is coming from the second-story eaves.*
- *Water supply is close to structure with two 1,500 gpm (5 680 L/min) hydrants available.*
- *There is an exposure on Delta side.*
- *Probabilities*
- *Occupant may be trapped on second floor.*

- *Evidence of potential for fire and smoke spread to the second floor.*
- *Potential for fire spread above and below entry access point on Alpha side.*
- *Possible access through patio door on Charlie side of lower (garage) level.*

2. Incident Priorities

What are your considerations for addressing the incident priorities of life safety, incident stabilization, and property conservation?

Life Safety

- *Report by owner indicates need for search and rescue.*

Incident Stabilization

- *Rapid fire attack required to support search and rescue team and prevent extension of the fire into upper story and attic area.*

Property Conservation

- *Rapid fire attack will reduce the potential for damage to the structure and contents.*

3. Operational Strategy

What are your considerations when selecting your operational strategy?

- *Your strategy choice may depend on how many resources you will have quickly available at the scene. There is a direct route into the fire area through the front door, which would make an offensive operation straightforward. If an offensive strategy is selected, adequate resources must be available to search the structure, remove the victim, and quickly confine and extinguish the fire in the room of origin, as well as addressing any fire extension to the garage, second story, and attic.*
- *A water supply is close to the structure that should be adequate to support an offensive attack or a defensive strategy for exposure protection.*

4. Command Options

What are your considerations when selecting a Command option: Investigation, Fast Attack, or Command?

- *Depending on the resources available, the Fast Attack option may initially be the most appropriate due to the need for search and rescue.*
- *Once sufficient resources are on scene, the Command option may be initiated.*

5. Tactics

What are your considerations, if any, for the tactics (RECEO-VS) used to accomplish your incident priorities?

Rescue

- *The integrity of the structure must be considered when conducting a primary and secondary search. There may be a collapse hazard from the floor and stairs leading to the second level due to fire extension. The placement of ground ladders at the windows of the second floor should be considered for both access and emergency escape by firefighters.*

- *Attack hoselines must also be in place to protect firefighters from fire extension above and below the search area.*
- *If the victim is overcome with smoke, additional personnel may be required to remove the victim.*

Exposure

- *The primary exposure considerations are the interior areas of the structure. Fire must be prevented from extending into the ground floor and the second floor as well as the attic.*
- *Attack hoselines must also be in place to prevent fire extension into these areas.*

Confinement

- *The initial attack lines through the front door, or the back door if necessary, are used to confine the fire to the room of origin.*
- *Access may also be gained through the garage door if time and resources permit.*

Extinguishment

- *Rapid fire attack and extinguishment are required to protect the search and rescue teams and the victim believed to be above the fire.*
- *Any crews working inside, especially for initial confinement and extinguishment, must maintain situational awareness of the potential for fire extension above and below them.*
- *Consideration must be given to quickly accessing the attic area during extinguishment to check for fire extension and, if necessary, to quickly extinguish any fire. Care must be taken to avoid or limit crews advancing hoselines into areas on the ground floor while there is active fire in the attic above them.*

Overhaul

- *During overhaul, consideration must be given to the structural integrity. Fire may have damaged the floor deck and/or joists of the second floor. It may be necessary to limit activities in any areas where there is a potential collapse.*
- *Walls must be opened to determine fire extension to the second floor and attic.*

Ventilation

- *Horizontal ventilation must be coordinated with the fire attack crews to prevent spreading or intensifying the fire.*
- *Once the fire is extinguished, mechanical horizontal ventilation can be used to remove the smoke from the structure.*

Salvage

- *Salvage covers and floor runners should be used to prevent additional damage to contents and floors.*
- *Properly applied ventilation techniques can reduce smoke damage.*

6. Unusual Hazards or Conditions

What are any unusual hazards or conditions which must be addressed? How will you address those hazards and/or conditions?

- *Multiple levels accessed through the front door create a hazard to firefighters by placing search teams above the fire, causing personnel to become disoriented, and complicating control and communication by an IC located outside the structure.*

- *Potential for fire extension into the attic on both the ground and second-story levels.*

Notes

Notes

Scenario 6

On Tuesday, July 6, at 5:45 p.m., dispatch reports a charcoal grill on fire on a second-floor balcony at the Hillside Village Apartments located at 816 E. Virginia. The dispatcher states that the caller reported heavy fire and black smoke coming from the balcony. The current weather condition is clear.

On Arrival
You are the first-arriving officer. You see fire on a second-floor balcony on the Alpha side of a multifamily apartment complex.

Water Supply
One blue 1,500 gpm (5 680 L/min) hydrant is located ½ block West on E. Virginia

Weather
- Temperature — 98°F (37°C)
- Wind direction — North
- Wind velocity — 5 mph (8 km/h)
- Humidity — 75%

360° View

Alpha Side

Bravo Side

Delta Side

Charlie Side

Chapter 6 • Residential Scenarios **231**

Site Plan

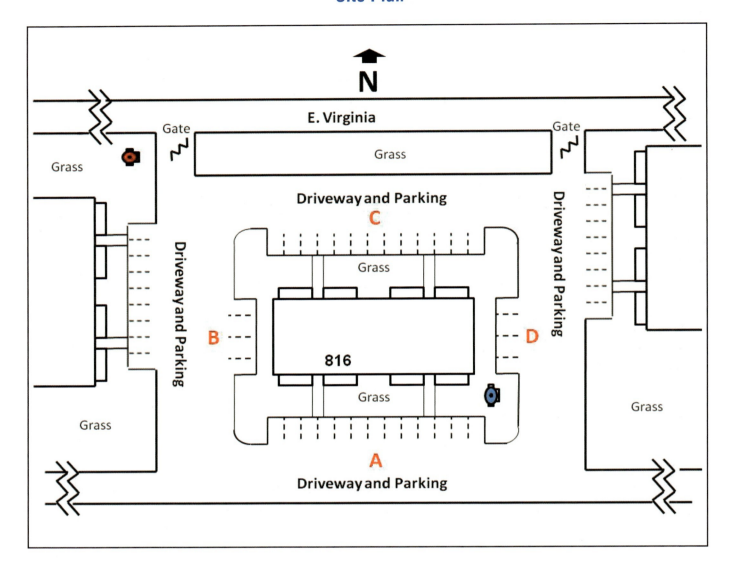

1. Size-Up

Based on the dispatch information and your visual observation of the scene on arrival, what are the facts and probabilities for this scenario?

Facts

- *Type V wood frame construction.*
- *Multifamily two-story residence that is occupied.*
- *Structure is set back from the parking lot.*
- *Open stairwell to second floor is adjacent to the fire.*
- *Heavy fire and smoke is visible on the balcony on the Alpha side.*
- *Water supply is close to structure with one 1,500 gpm (5 680 L/min) hydrant available.*

Probabilities

- *High life-safety risk.*
- *Structure may contain as many as 16 units divided into 8 unit blocks by interior fire wall.*
- *Fire may have extended into apartment and through eaves into attic.*
- *Lightweight construction may promote rapid fire spread into common attic space.*
- *Water pressure may be affected by limited hydrant coverage.*

2. Incident Priorities

What are your considerations for addressing the incident priorities of life safety, incident stabilization, and property conservation?

Life Safety

- *Time of day increases the possibility that most of the units are occupied.*
- *The primary priority is search and rescue of occupants in unit attached to the balcony on fire and evacuation of all other residents.*

Incident Stabilization

- *Rapid fire extinguishment is needed to prevent fire extension into the apartment and into the attic.*

Property Conservation

- *Rapid fire extinguishment will decrease the possibility of damage to the structure and contents of the adjacent apartment.*

3. Operational Strategy

What are your considerations when selecting your operational strategy?

- *An offensive strategy is required to prevent loss of life and to reduce property damage.*

4. Command Options

What are your considerations when selecting a Command option: Investigation, Fast Attack, or Command?

- *Depending on the resources available, the initial Command option employed is Fast Attack. All personnel are required for search and rescue and fire extinguishment.*
- *When resources permit, implementation of the "Command" option should occur. The size of the fire scene, the number of units involved, and the potential for an increase in the size and scope of the fire require the IC to coordinate activities from the Command Post.*

5. Tactics

What are your considerations, if any, for the tactics (RECEO-VS) used to accomplish your incident priorities?

Rescue

- *Search and rescue of the apartment next to the fire can occur as the initial attack hoseline is advanced into the apartment to extinguish the fire from that side.*
- *Search, rescue, and evacuation of the remaining units starts with those closest to the fire.*

Exposure

- *The primary exposure is the adjacent apartment and attic. The adjacent apartment is protected by the initial attack hoseline.*
- *A second hoseline can be directed from the ground at the eave to prevent or reduce the effects of the fire upon it and control extension into the attic.*

Confinement

- *Placement of the initial and secondary hoselines can confine the fire to the balcony area.*

Extinguishment

- *Rapid deployment of the initial attack line through the apartment can extinguish the fire.*

Overhaul

- *Once the fire is extinguished, opening the walls and balcony ceiling is required to determine the extent of fire extension and extinguish it.*
- *Gaining access to the attic to determine extension and extinguish any fires in that area.*
- *Placement of ground ladders under the eaves to open them and extinguish any fire spread.*

Ventilation

- *Horizontal ventilation is used for the apartment which is adjacent to the fire.*
- *Other apartments are ventilated using the most appropriate means.*

Salvage

- *Salvage covers and floor runners are deployed in the apartment adjacent to the fire.*
- *Furnishings in the apartment below are moved or covered to prevent water damage.*
- *Excess water from the balcony and apartment should be removed to prevent structural damage to the balcony.*

6. Unusual Hazards or Conditions

What are any unusual hazards or conditions which must be addressed? How will you address those hazards and/or conditions?

- *Potential for fire spread under the eave and behind the exterior siding into the attic.*

- *Search, rescue, and evacuation will require additional resources.*
- *Large open area in attic and exposed wood trusses create a high risk of fire extension.*
- *Weather conditions can affect the physical condition of firefighters due to the heat and humidity. Medical monitoring, rehabilitation, and relief personnel may be rapidly needed.*

Notes

Notes

Scenario 7

On Tuesday, March 4, at 2:45 p.m., dispatch reports a fire alarm activation at The Links Apartments, building 2300 located on W. Adams Street. The dispatcher states that the fire suppression system has been activated. The current weather condition is clear.

On Arrival
You are the first-arriving officer. You see a two-story, multifamily apartment complex with light smoke coming from a bottom floor window on the Alpha side of the structure.

Water Supply
One blue 1,500 gpm (5 680 L/min) hydrant is located directly South across W. Adams Street.

One red 450 gpm (1 900 L/min) hydrant is located in the Southwest corner of the parking lot.

Weather
- Temperature — 68° F (20°C)
- Wind direction — South
- Wind velocity — 5 mph (8 km/h)
- Humidity — 58%

360° View

Alpha Side

Bravo Side

Charlie Side

Delta Side

Site Plan

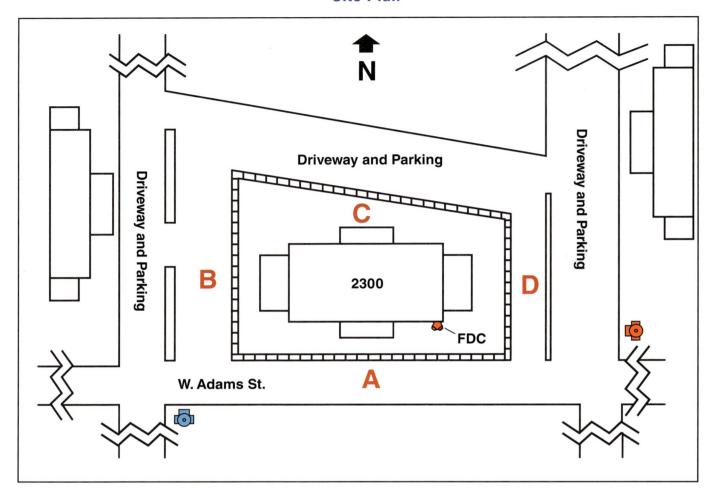

1. Size-Up

Based on the dispatch information and your visual observation of the scene on arrival, what are the facts and probabilities for this scenario?

Facts

- Type V wood frame construction with masonry veneer.
- Multifamily residence containing eight units that is occupied.
- Structure is equipped with a fire suppression system that is supported by a FDC located on Alpha side near the Delta corner.
- Light smoke is visible coming from a window on the ground floor on the Alpha side.
- Water supply is close to structure with one 1,500 gpm (5 680 L/min) hydrant and one 450 gpm (1 900 L/min) hydrant available.
- An audible water flow alarm has been activated.

Probabilities

- Based on the dispatcher's report and the water flow alarm, it is likely that the fire suppression system has been activated.

- *The color and quantity of the visible smoke indicates that a fire has been or is being confined by the fire suppression system.*
- *It is possible that at least one of the units is occupied due to the presence of a car parked on the Alpha side.*
- *There appear to be internal exposures above and on the Charlie and Delta sides of the unit.*

2. Incident Priorities

What are your considerations for addressing the incident priorities of life safety, incident stabilization, and property conservation?

Life Safety

- *Initial investigation must be performed to determine if occupants are in immediate danger in the primary apartment and those adjacent to it.*

Incident Stabilization

- *Due to the activation of the fire suppression system and the presence of light smoke, an attack line should be placed in operation at the door to the apartment.*
- *Depending on local SOP/SOG, hoselines should be connected to the FDC and charged to support the fire suppression system. Either the first or second arriving unit should be assigned this task.*
- *To reduce further water damage, the suppression system must be shut off as soon as it is determined the fire has been extinguished.*

Property Conservation

- *To reduce further water damage, the suppression system must be shut off as soon as it is determined the fire has been extinguished.*
- *Horizontal ventilation can be used to reduce further smoke damage.*

3. Operational Strategy

What are your considerations when selecting your operational strategy?

- *An offensive strategy is justified by the visual indicators and dispatch report.*

4. Command Options

What are your considerations when selecting a Command option: Investigation, Fast Attack, or Command?

- *The most appropriate Command option is Investigation. However, a hoseline should be ready for operation if the fire has extended into areas that are not protected by the suppression system.*

5. Tactics

What are your considerations, if any, for the tactics (RECEO-VS) used to accomplish your incident priorities?

Rescue

- *Search, rescue, and evacuation are performed during the investigation of the apartment.*

Exposure
- *Exposures are controlled by the structure and the fire suppression system.*

Confinement
- *The fire is confined by the structure design and the fire suppression system.*

Extinguishment
- *Extinguishment is accomplished by supporting the fire suppression system.*
- *If needed, an attack line can be used to complete the task.*

Overhaul
- *Removal of smoldering debris and opening of any walls or ceiling to locate any fire extension.*

Ventilation
- *Horizontal ventilation is used to remove smoke from the apartment.*

Salvage
- *The fire suppression system is shut off when the fire is determined to be controlled reducing further water damage.*
- *Smoke is removed with horizontal ventilation.*
- *Water is removed along with debris.*

6. Unusual Hazards or Conditions

What are any unusual hazards or conditions which must be addressed? How will you address those hazards and/or conditions?

- *No unusual hazards in this scenario.*

Scenario 8

On Saturday, November 20, at 10:45 p.m., dispatch reports a structure fire at the Boomer Village Apartments, building 5840 located on N. Boomer Road. The current weather condition is light rain.

On Arrival
You are the first-arriving officer. You see a two-story, multifamily apartment complex with heavy smoke and fire showing from the attic vents on Bravo and Delta Sides of the structure.

Water Supply
One Red 500 gpm (1 900 L/min) hydrant is located 1 block South on N. Boomer Road.

Weather
- Temperature — 34°F (1°C)
- Wind direction — North
- Wind velocity — 5 mph (8 km/h)
- Humidity – 100%

360° View

Alpha Side

Bravo Side

Charlie Side

Delta Side

Site Plan

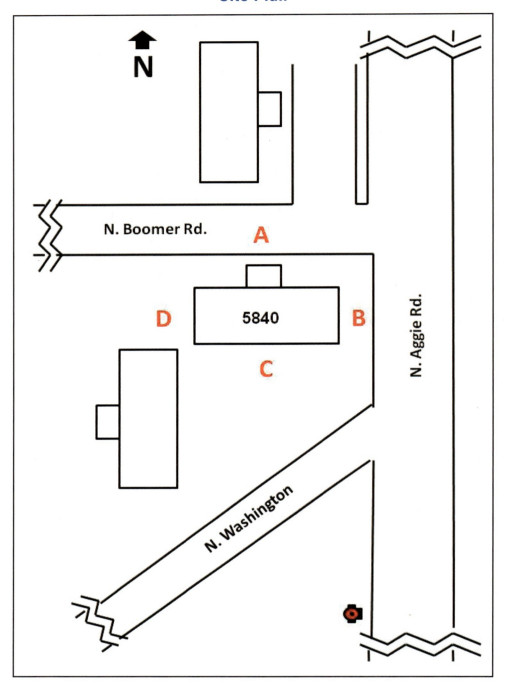

1. Size-Up

Based on the dispatch information and your visual observation of the scene on arrival, what are the facts and probabilities for this scenario?

Facts

- *Type V construction with brick veneer.*
- *Occupied multifamily two-story residence.*
- *Structure contains four units accessed by a central hallway and stairs and doors on the Alpha and Charlie sides.*

- *Smoke and fire are coming from the peak on the Bravo and Delta sides.*
- *Water supply is close to structure with one 500 gpm (1 900 L/min) hydrant available.*
- *There is an exposure on the Charlie/Delta corner.*
- *There is a gas service meter on the Charlie side.*

Probabilities

- *The time of night indicates that occupants may be asleep.*
- *Search, rescue, and evacuation will require high levels of resources.*
- *There is utility service in the building including electricity and gas.*
- *Fire maybe concentrated in the attic and have weakened the roof structure.*

2. Incident Priorities

What are your considerations for addressing the incident priorities of life safety, incident stabilization, and property conservation?

Life Safety

- *Time of night and type of occupancy increase the risk to occupant lives.*
- *The location of the visible fire and lack of visible occupants indicate that occupants may not be aware of the danger to them.*

Incident Stabilization

- *Rapid fire extinguishment to prevent further damage to the structure and threat to occupants.*

Property Conservation

- *The use of salvage covers and floor runners to reduce or prevent water damage and damage from any falling debris.*
- *Horizontal ventilation to reduce damage to the structure and contents.*

3. Operational Strategy

What are your considerations when selecting your operational strategy?

- *An offensive strategy is required to evacuate occupants and rapidly locate and extinguish the fire.*

4. Command Options

What are your considerations when selecting a Command option: Investigation, Fast Attack, or Command?

- *Depending on the resources available, the initial Command option employed is Command. The size of the fire, the time of night, potential for search and rescue, the number of responding units involved, the limited means of egress, the need for coordination of units and activities, and the potential for an increase in the size and scope of the fire require the IC to coordinate activities from the Command Post.*

5. Tactics

What are your considerations, if any, for the tactics (RECEO-VS) used to accomplish your incident priorities?

Rescue
- *Initial activities must include search and evacuation of the entire structure.*
- *Due to the configuration of the structure, a charged hoseline must be in place in the central hallway in the event that fire is located in one of the apartments during the search operation.*

Exposure
- *The exterior exposure on the Charlie/Delta corner is not a concern.*
- *Internal exposures include the egress stairs and hallway and other apartments in the structure.*

Confinement
- *Once located, the fire must be confined to the area of origin.*
- *The initial attack line in the central hallway can be used for this purpose.*

Extinguishment
- *The initial attack hoseline in the central hallway is used to extinguish any fire in an apartment.*
- *If the fire is located in the attic space, open the ceiling on the 2nd floor and direct a fire stream into the attic space.*
- *Location and size of the fire as well as the weather conditions would prohibit placing personnel on the roof.*

Overhaul
- *Walls and ceiling must be opened to locate the avenues of fire spread and any additional fire in those cavities.*

Ventilation
- *Horizontal ventilation is used in any smoke-filled apartments.*
- *Location and size of the fire as well as the weather conditions would prohibit placing personnel on the roof to perform vertical ventilation.*

Salvage
- *Contents in all apartments must be covered with salvage covers to prevent water damage.*
- *Horizontal ventilation is used to remove smoke and reduce smoke damage.*

6. Unusual Hazards or Conditions

What are any unusual hazards or conditions which must be addressed? How will you address those hazards and/or conditions?

- *Occupancy type and time of night increase the need for additional resources for search, rescue, and evacuation.*
- *Light rain and near freezing temperatures increase the hazards of working on the roof of the structure.*
- *Initial visual indications that the fire is well involved, located in the attic, and difficult to access.*

Scenario 9

On Saturday, August 18, at 9:00 a.m., dispatch reports a structure fire located 1256 South Stone Ridge Drive. The dispatcher states that the alarm company reported the incident. Current weather condition is clear.

On Arrival
You are the first-arriving officer. You see a large two-story residential structure with heavy fire and smoke showing from the roof.

Water Supply
One blue 1,500 gpm (5 680 L/min) hydrant is located one block North

One blue 1,500 gpm (5 680 L/min) hydrant is located one block South

Weather
- Temperature — 72°F (23°C)
- Wind direction — South
- Wind velocity — 10 mph (16 km/h)
- Humidity — 78%

360° View

Alpha Side

Bravo Side

Charlie Side

Delta Side

Chapter 6 • Residential Scenarios

Site Plan

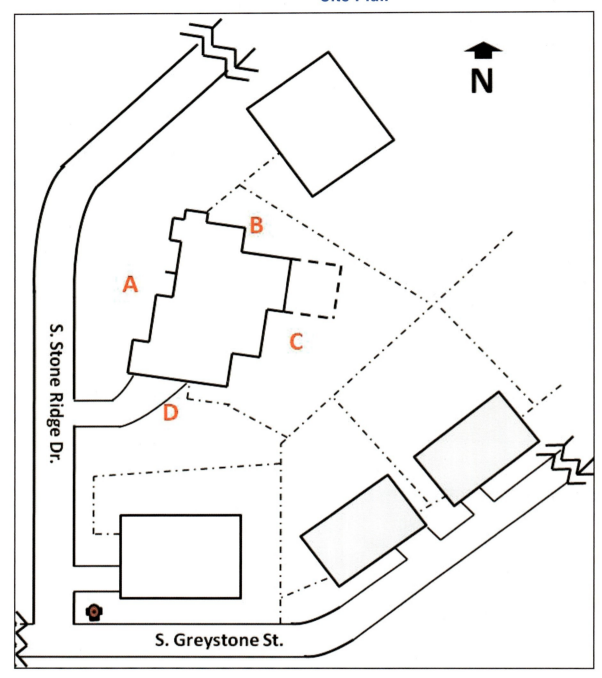

1. Size-Up

Based on the dispatch information and your visual observation of the scene on arrival, what are the facts and probabilities for this scenario?

Facts

- *Type V wood frame construction with brick veneer.*
- *Lightweight truss construction.*
- *General floor arrangement typical of this type of structure.*
- *High ceilings in part of living area.*

- *Steep roof.*
- *Sleeping rooms on upper floor.*
- *HVAC units located in attic.*
- *Single family two-story residence.*
- *Fire is visible coming from the roof on all sides.*
- *Water supply is close to structure with two 1,500 gpm (5 680 L/min) hydrants available.*
- *There are no exposures near the structure.*
- *Access on the Charlie side is limited due to the lot arrangement and privacy fences.*
- *Large porch on the Charlie side is a potential collapse hazard.*

Probabilities
- *Occupants may be at home due to the time of day.*
- *Rapid internal fire spread.*
- *Structural weakness due to use of lightweight construction and volume of fire.*
- *HVAC in attic and large chandelier in entry may fall or cause ceiling to collapse when weakened by fire and water.*
- *Once affected by fire, the porch on the Charlie side may collapse.*

2. Incident Priorities

What are your considerations for addressing the incident priorities of life safety, incident stabilization, and property conservation?

Life Safety
- *Time of day and day of week may increase the risk to occupant lives.*
- *The location of the visible fire and lack of visible occupants indicates that occupants may not be aware of the danger to them.*
- *The location of the fire near the bedrooms on the second floor will require firefighters to search areas that may be involved in fire.*

Incident Stabilization
- *Volume and location of fire will require a rapid fire attack from the exterior using large diameter nozzles or master streams.*
- *Interior attack lines will be required to protect search and rescue teams.*

Property Conservation
- *If resources are available, salvage covers should be placed over items on the first floor.*
- *Property conservation may not be possible given the volume and location of the fire.*

3. Operational Strategy

What are your considerations when selecting your operational strategy?

- *An offensive strategy is required to locate and rescue occupants.*
- *Due to the volume and location of the fire, a transition to a defensive strategy may be necessary once the rescue tactic is completed and all firefighters have withdrawn from the structure.*

4. Command Options

What are your considerations when selecting a Command option: Investigation, Fast Attack, or Command?

- *Depending on the resources available, the initial Command option employed is Command. The size of the fire, the potential for search and rescue, and the potential for structural collapse demand that the IC have overall visual control of the incident.*

5. Tactics

What are your considerations, if any, for the tactics (RECEO-VS) used to accomplish your incident priorities?

Rescue

- *Initial activities must include search and rescue of the occupants.*
- *Due to the configuration of the structure, a charged hoseline must be in place in the entry to protect search teams during the search and rescue operation.*
- *Crews working on the second floor must be cautious of a ceiling collapse. The volume of fire indicates fast fire growth in the attic – maybe a broken or leaking gas line. Remember, except for the trusses and the roof decking, there is not much fuel loading in an attic, certainly not enough to cause this amount of fire this quickly on a Saturday morning in a house that has a monitored detection system.*

Exposure

- *Wood shingle roofs on adjacent structures constitute exterior exposures. Additional resources must be assigned to spark patrol in adjacent areas.*
- *Internal exposures include the stairs to the second floor and any unburned areas of the structure.*

Confinement

- *Fire should be confined to the roof and attic area of the structure and prevented from extending into the second floor.*

Extinguishment

- *Due to the volume and location of the fire, the application of fire streams from the ground and from elevated master streams on the fire should be considered.*

Overhaul

- *Once the fire is extinguished, overhaul operations in the remainder of the structure can occur.*

Ventilation
- *Due to the volume and location of the fire, ventilation will not be required.*

Salvage
- *Due to the volume and location of the fire, salvage will be limited to covering or removing items from the ground floor, if resources are available.*

6. Unusual Hazards or Conditions
What are any unusual hazards or conditions which must be addressed?
- *How will you address those hazards and/or conditions?*
- *High ceilings in entry and living area.*
- *Lightweight truss construction.*
- *Steep roof.*
- *Large mass of fire.*
- *Heavy objects, such as HVAC systems and chandeliers, pose a collapse hazard.*
- *Large area and multiple rooms require more time for search and rescue.*

Notes

Commercial Scenarios

Chapter Contents

Scenario 1 259	Scenario 4 281
Scenario 2 267	Scenario 5 287
Scenario 3 273	Scenario 6 295

Divider page photo courtesy of Chris Mickal, New Orleans (LA) FD .

chapter 7

Scenarios

This chapter contains six hypothetical scenarios based on commercial occupancy types that are found in most response areas. The scenarios contain best practices that can be applied to each type of situation. However, these are suggestions only and not hard and fast rules that must be applied in every incident.

Because these scenarios are examples to help you learn how to make decisions based on the information in the previous chapters, there are limitations placed on them. The resources available to you will arrive within 10 minutes. You are the Incident Commander (IC) and must allocate your resources to control/mitigate the incident to the best of your ability. The resources you have reflect those in your local department, including first-alarm assignments, mutual aid, equipment, and staffing.

Commercial Scenarios

Learning Objectives

1. Given information on a commercial fire and resources available to your department, develop an Incident Action Plan (IAP) for achieving the incident objectives.

Note to Instructors

The following scenarios contain recommended practices based on the facts provided and on information contained in the text of this manual. These scenarios are intended to be used as teaching aids. It is suggested that you develop scenarios specific to your department's resources, procedures, and response area. Your scenarios should contain structures, hazards, water supplies, photographs, and site plans that exist in your jurisdiction. Do not include probabilities, incident priorities, strategies, command options, or tactics. This information should be supplied by the students based on their understanding of the text. Your scenarios can be varied based on changes to the water supply (hydrants out of service, peak water demand times), delayed resources, weather conditions, minimum staffing, time of day, or local standard operating procedures (SOPs).

Chapter 7
Commercial Scenarios

Scenario 1

On Friday, February 20, at 10:45 a.m., dispatch reports a structure fire at the Lakeside Strip Mall at 320 North Redbud Avenue. Current weather conditions are clear skies.

On Arrival
You are the first-arriving officer. You see a single-story small strip shopping center with heavy smoke and fire showing from a unit on the west end of the Alpha side of the structure.

Water Supply
One red 500 gpm (1 900 L/min) hydrant directly in front of the structure on E. Redbud Avenue

One red 500 gpm (1 900 L/min) hydrant ½ block West on E. Redbud Avenue

Weather
- Temperature – 45°F (7°C)
- Wind direction – South
- Wind velocity – 10 mph (16.1 km/h)
- Humidity – 48%

360° View

Bravo Side

Alpha Side

Delta Side

Chapter 7 • Commercial Scenarios **261**

Site Plan

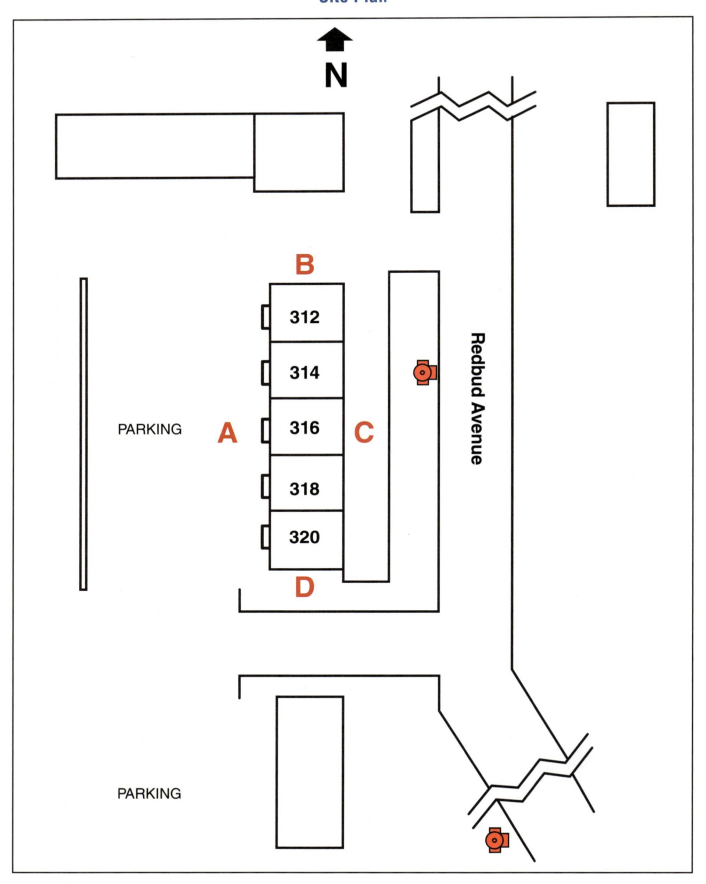

1. Size-Up

Based on the dispatch information and your visual observation of the scene on arrival, what are the facts and probabilities for this scenario?

Facts

- *Type II construction.*
- *Commercial occupancy.*
- *Multiple units that may or may not be separated by fire-rated walls.*
- *There appears to be a continuous cockloft.*
- *Fire and smoke are visible inside the unit on the Delta end of the mall.*
- *Water supply is close to structure with two 500 gpm (1 900 L/min) hydrants available.*
- *The adjacent unit on the Delta side is the closest exposure, Unit #318.*

Probabilities

- *Lack of cars during a normal business day indicates that the units may be unoccupied or closed.*
- *Steel roof trusses may have been exposed to fire for an unknown period of time creating a collapse hazard.*
- *Interior wall on Bravo side of the involved unit may provide a barrier to fire extension.*
- *No fire or smoke is visible in units on the Bravo side of involved unit.*

2. Incident Priorities

What are your considerations for addressing the incident priorities of life safety, incident stabilization, and property conservation?

Life Safety

- *Volume of visible fire in Unit 320 indicates that savable lives may not be present.*
- *Search of units 312 through 318 for possible victims, beginning with Unit 318.*

Incident Stabilization

- *Rapid fire attack from the Alpha side for quick extinguishment.*

Property Conservation

- *Volume of fire in Unit 320 would indicate that contents are beyond saving.*
- *Salvage covers for contents in Units 312 through 318 to protect from water, smoke, or debris.*

3. Operational Strategy

What are your considerations when selecting your operational strategy?

- *Your strategy choice may depend on how many resources you will have quickly available at the scene. There is a direct route into the fire area through the front of the unit, which would make an offensive operation straightforward. If an offensive strategy is selected, adequate resources must be available to quickly confine and extinguish the fire to prevent any fire extension to the cockloft or through wall penetrations.*

- *If there is an obvious structural collapse concern, a defensive strategy would be appropriate. Establish a collapse zone on Alpha, Charlie, and Delta sides. Monitor the condition of the overhang on the Alpha side.*

4. Command Options

What are your considerations when selecting a Command option: Investigation, Fast Attack, or Command?

- *Due to the volume of fire, size of the structure, and resources available, the Command option may be the most appropriate.*

5. Tactics

What are your considerations, if any, for the tactics (RECEO-VS) used to accomplish your incident priorities?

Rescue

- *Perform a primary search of Units 312 through 318, beginning with 318.*

Exposure

- *Monitor conditions in Unit 318 by removing any ceiling tiles and using a hoseline to protect any penetrations of the wall.*
- *Monitor the condition of the roof for sagging and potential collapse.*

Confinement

- *Use attack hoselines to confine the fire to the unit of origin.*
- *Monitor the condition of the wall on the interior Bravo side for deterioration from heat or roof collapse.*

Extinguishment

- *An offensive or defensive attack through the Alpha side opening to extinguish the fire.*
- *Remember that the walls are masonry and will retain heat in the compartment.*

Overhaul

- *Once the fire is controlled, pull down remaining ceiling tiles, open any walls, and inspect Bravo-side units for fire extension.*

Ventilation

- *Due to the potential for roof collapse, cross ventilation should be used to remove heat and smoke.*

Salvage

- *Volume of fire would indicate that contents are beyond saving.*
- *Use salvage covers in Units 312 through 318 to protect any contents from smoke, water, and debris.*
- *Monitor condition of Alpha side overhang, exterior walls, and roof for possible collapse.*

6. Unusual Hazards or Conditions
What are any unusual hazards or conditions which must be addressed? How will you address those hazards and/or conditions?

- *Lightweight construction overhang on Alpha side.*
- *Unknown condition of interior Bravo-side wall including possible penetrations.*
- *Possible common attic and exposed metal roof trusses.*
- *Retaining wall and narrow access on Charlie side.*

Notes

Notes

Scenario 2

On Monday, May 20, at 2:40 p.m., dispatch reports a structure fire at the Town and Country Insurance building located at 1100 S. Country Club road. Dispatch reports that the fire alarm has been activated on the second floor. Current weather conditions are clear skies.

On Arrival

You are the first-arriving officer. You see a two-story commercial office building with fire showing through a second-story window from the Bravo side of the structure.

Water Supply

One blue 1,500 (5 680 L/min) hydrant located at the northeast corner of the parking lot

One blue 1,500 (5 680 L/min) hydrant located one block West on W. Frontage Rd.

Weather

- Temperature – 68°F (20°C)
- Wind direction – North
- Wind velocity – 15 mph (24 km/h)
- Humidity – 85%

360° View

Alpha Side

Bravo Side

Charlie Side

Delta Side

Chapter 7 • Commercial Scenarios

Site Plan

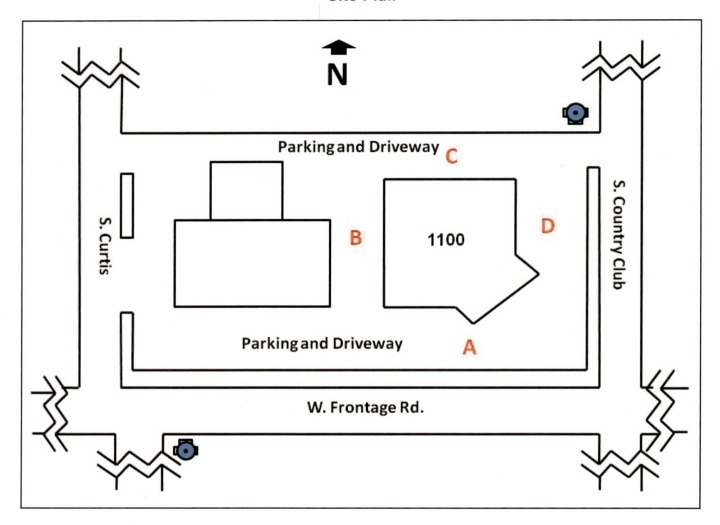

1. Size-Up

Based on the dispatch information and your visual observation of the scene on arrival, what are the facts and probabilities for this scenario?

Facts

- *Type V construction.*
- *Business occupancy containing offices and low to moderate fuel load.*
- *Fire alarm on the second floor has been activated.*
- *Smoke is coming from a window on Bravo side, second floor.*
- *Water supply is close to structure with two 1,500 (5 680 L/min) hydrants available.*
- *There is an exposure on Bravo side.*
- *The time of day and presence of cars in the parking lot would indicate that the structure is occupied.*
- *An atrium open to both floors is visible on the Alpha side.*
- *Multiple electrical and gas connections for each office unit.*

Probabilities
- *The location of the smoke and the activated alarm make it possible that fire could extend into the attic area.*
- *The type of construction would indicate unprotected metal roof trusses in the attic.*
- *The structure size and occupancy type would indicate a lack of any fire suppression system.*
- *Lack of occupants in the front of the structure would indicate that the local alarm had not alerted them to the hazard, they are unaware of the hazard, or are attempting to control the hazard themselves.*

2. Incident Priorities

What are your considerations for addressing the incident priorities of life safety, incident stabilization, and property conservation?

Life Safety
- *Time of day, occupancy type, and lack of self-evacuated occupants would indicate the need for search, rescue, and evacuation.*
- *Search activities can be combined with fire attack on the second floor.*

Incident Stabilization
- *Rapid fire attack and extinguishment can accomplish all three incident priorities by confining the fire to the area of origin.*
- *Initially, the exterior exposure on the Bravo side is not a consideration.*

Property Conservation
- *Rapid application of water on the fire can reduce potential fire and smoke damage as well as limit water damage.*
- *Proper ventilation techniques can quickly remove smoke from the structure.*
- *Proper salvage techniques can protect contents from smoke, water, and debris damage.*

3. Operational Strategy

What are your considerations when selecting your operational strategy?
- *An offensive strategy is appropriate based on the limited amount of visible fire and smoke, the need for evacuation, and the need for rapid fire extinguishment.*
- *Resources are available to safely support an offensive strategy.*

4. Command Options

What are your considerations when selecting a Command option: Investigation, Fast Attack, or Command?
- *Depending on the available resources, the Fast Attack option is the most appropriate.*
- *Search, rescue, and evacuation as well as fire attack will require high commitment of resources.*
- *If resources are not available, immediate extinguishment of the fire can accomplish the incident priorities while occupants self-evacuate.*

5. Tactics

What are your considerations, if any, for the tactics (RECEO-VS) used to accomplish your incident priorities?

Rescue

- *Search and rescue of occupants in the vicinity of the fire and evacuation of the rest of the occupants is the primary mission.*
- *This tactic can be combined with the advancement of fire attack hoselines.*

Exposure

- *Interior exposures, the adjacent compartments, and the attic are the primary concern.*
- *Rapid fire extinguishment will protect them until overhaul can locate any fire extensions.*

Confinement

- *Rapid fire attack will confine the fire to the area of origin.*

Extinguishment

- *The apparent small volume of fire would indicate that a rapid extinguishment is possible.*
- *Rapid extinguishment will protect occupants who are not close to the fire, limit spread, and reduce damage.*

Overhaul

- *Ceiling tiles and walls should be opened to determine the extent of fire spread.*

Ventilation

- *Horizontal ventilation should be used to clear the second floor.*
- *Vertical ventilation is only recommended if the fire has extended into the attic.*
- *Use caution when working on the roof to ensure that the trusses have not weakened.*

Salvage

- *Salvage covers and floor runners should be used on the first floor beneath the fire area and in areas adjacent to the fire.*
- *Proper ventilation techniques should be applied quickly, in coordination with the attack team, to limit smoke damage.*

6. Unusual Hazards or Conditions

What are any unusual hazards or conditions which must be addressed? How will you address those hazards and/or conditions?

- *Coordinating evacuation of building occupants.*
- *Unknown extent of fire.*
- *Multiple possibilities of fire spread.*
- *Separate utility meters that may make it difficult to isolate utilities.*
- *Traffic control.*
- *Lightweight building construction.*

Scenario 3

On Monday, December 20, at 9:30 a.m., dispatch reports a structure fire at the Urgent Care Clinic located at 2300 Nebraska Street. Dispatch reports that the fire alarm has been activated in the lobby area.

On Arrival
You are the first-arriving officer. You see a single-story commercial building with smoke and fire showing from the Alpha side of the structure.

Water Supply
One red 500 gpm (1 900 L/min) hydrant on the southeast corner of parking lot

Weather
- Temperature – 25°F (4°C)
- Wind direction – North
- Wind velocity – 5 mph (8 km/h)
- Humidity – 80%

360° View

Alpha Side

Bravo Side

Charlie Side

Delta Side

Chapter 7 • Commercial Scenarios 275

Site Plan

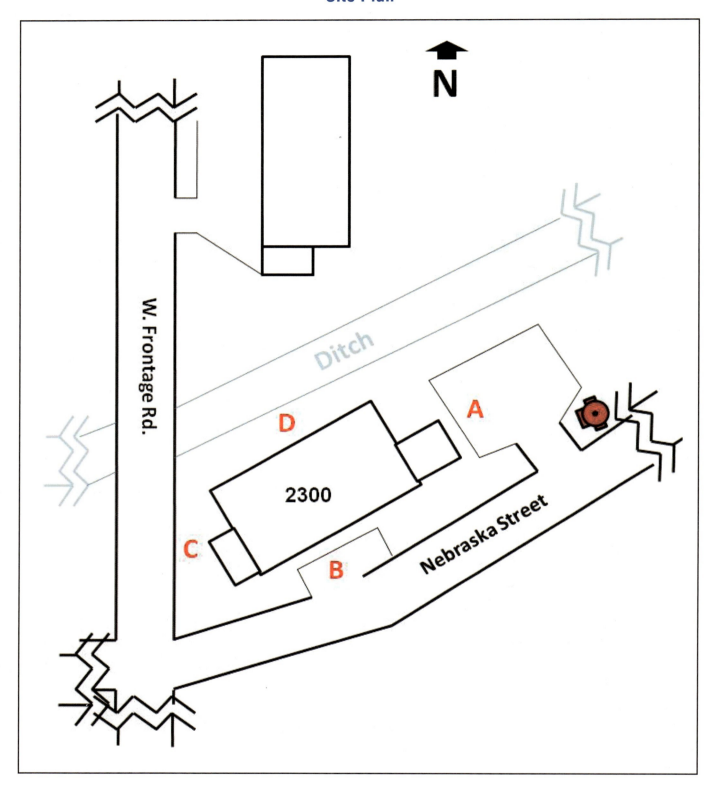

1. Size-Up

Based on the dispatch information and your visual observation of the scene on arrival, what are the facts and probabilities for this scenario?

Facts

- *Type V wood frame construction with brick veneer.*
- *Business occupancy that is occupied.*
- *Fire is visible in a front door on the Alpha side.*
- *Water supply is close to structure with one 500 gpm (1 900 L/min) hydrant available.*
- *Access to the Delta side is obstructed by a ditch and an overhead power line.*
- *A second access/egress door is on the Bravo side.*
- *There is a large-volume attic space above the main floor.*
- *Compressed gas cylinders, including medical oxygen.*
- *Biological waste.*

Probabilities

- *Time of day and occupancy type indicate that the structure may be occupied.*
- *Occupants are not visible, indicating they are either attempting to control the fire or trapped by it.*
- *No smoke is visible, indicating that the fire is confined to the interior of the building.*
- *Some occupants may be nonambulatory.*

2. Incident Priorities

What are your considerations for addressing the incident priorities of life safety, incident stabilization, and property conservation?

Life Safety

- *Lack of visible occupants outside the structure indicate that search and rescue are required initially.*
- *Access through the door on Bravo side may be the most direct route.*

Incident Stabilization

- *Rapid fire extinguishment will resolve all incident priorities.*
- *Access through the door on Bravo side will help in search and rescue and also place an attack hoseline between the occupants and the fire.*

Property Conservation

- *Rapid fire extinguishment will reduce structural and content damage.*
- *Application of proper salvage and ventilation techniques will limit property damage.*

3. Operational Strategy

What are your considerations when selecting your operational strategy?

- *An offensive strategy is required to locate and evacuate occupants and extinguish the fire.*

4. Command Options

What are your considerations when selecting a Command option: Investigation, Fast Attack, or Command?

- *The need for rapid action justifies the use of the Fast Attack option.*
- *Once additional resources arrive, the initial IC should transfer Command and switch to Command mode operating from the Command Post.*

5. Tactics

What are your considerations, if any, for the tactics (RECEO-VS) used to accomplish your incident priorities?

Rescue

- *Search, rescue, and evacuation of occupants should be accomplished initially.*
- *The access door on the Bravo side should be used for this tactic and for the advancement of a hoseline into the structure.*

Exposure

- *There is no external exposure requiring protection.*
- *The primary internal exposures are the offices adjacent to the lobby and the attic above the fire area.*
- *Placing an attack hose line between the lobby fire and the remainder of the interior will protect this exposure.*

Confinement

- *The fire can be confined to the lobby by taking the hose through the access door on the Bravo side.*

Extinguishment

- *A rapid attack from the unburned side of the interior will extinguish the fire with limited damage to structure and contents.*
- *If the fire is small in nature and limited in area, it may be possible to attack it from the lobby entrance door with a rapid application of water. Care must be taken not to drive the fire back into the unburned portions of the structure, into the attic (by blowing out ceiling tiles), or onto occupants or search teams.*

Overhaul

- *Ceiling tiles must be removed to determine fire extension into the attic.*
- *Smoldering debris should be removed from the lobby area.*
- *Walls and voids should be opened up to locate any remaining fire.*

Ventilation

- *Horizontal ventilation can be used through the lobby door to pull out smoke.*
- *Mechanical horizontal ventilation can be used through the access door on Bravo side if an exit point is available for the smoke closer to the lobby.*

Salvage

- *Application of proper ventilation techniques will reduce smoke damage.*
- *Use of salvage covers and floor runners will limit water and debris damage.*

6. Unusual Hazards or Conditions
What are any unusual hazards or conditions which must be addressed? How will you address those hazards and/or conditions?

- *Medical equipment (oxygen tank, biomedical waste).*
- *Potential roof collapse if fire gets into the attic area.*
- *Lightweight construction.*
- *Ambulatory/nonambulatory patients.*
- *Limited paths of egress.*
- *Overhead utility lines.*

Notes

Notes

Scenario 4

On Sunday, September 19, at 5:45 a.m., dispatch reports a structure fire at a restaurant located at 1500 N. Perkins Street. Dispatch reports that the building has been in the process of remodeling for the last three months. Current weather conditions are clear skies.

On Arrival
You are the first-arriving officer. You see a single-story commercial building with heavy smoke and fire showing from the roof of the structure.

Water Supply
One blue 1,500 gpm (5 680 L/min) hydrant at N. Perkins Street and the North Mall Access Road

Weather
- Temperature – 80°F (27°C)
- Wind direction – South
- Wind velocity – 10 mph (16 km/h)
- Humidity – 68%

360° View

Alpha Side

Bravo Side

Charlie Side

Delta Side

Chapter 7 • Commercial Scenarios   **283**

Site Plan

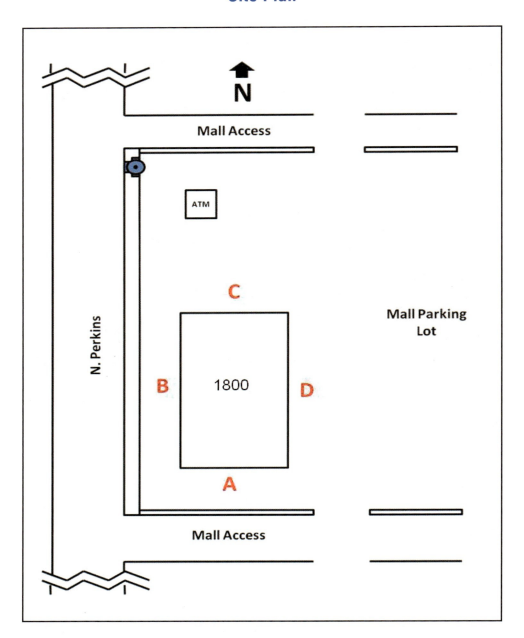

1. Size-Up

Based on the dispatch information and your visual observation of the scene on arrival, what are the facts and probabilities for this scenario?

Facts

- *Type V construction with stucco exterior.*
- *Assembly occupancy, unoccupied.*
- *Large amount of fire is visible from the roof near the Bravo side.*
- *Fire has caused the roof to collapse or open up.*
- *Water supply is close to structure with one 1,500 gpm (5 680 L/min) hydrant available.*

- *There are no exposures near the structure.*
- *Building is under renovation.*
- *One vehicle is present in the parking lot.*

Probabilities

- *Hazardous materials may be present and fire growth and extension may be accelerated due to the materials being used for renovation.*
- *Interior finish materials may be combustible.*
- *Opening in walls and ceiling may exist to permit rapid fire extension.*
- *Lightweight wood trusses in the attic are being exposed to heat and flames due to extension.*
- *Interior of structure may be fully involved.*
- *There is utility service that may include both electricity and gas.*

2. Incident Priorities

What are your considerations for addressing the incident priorities of life safety, incident stabilization, and property conservation?

Life Safety

- *The possibility that a savable victim is in the structure is reduced by the time of morning and the volume of fire.*
- *Firefighter life safety would take precedence justifying a defensive strategy.*

Incident Stabilization

- *The volume of fire visible and the apparent collapse of the roof would indicate the need for master stream appliances to extinguish the fire.*
- *If available, an elevated water stream is placed into the roof opening plus large volume fire streams are positioned through the doors and windows.*

Property Conservation

- *Property conservation may no longer be a priority due to the large volume of visible fire.*

3. Operational Strategy

What are your considerations when selecting your operational strategy?

- *The volume of fire and the questionable condition of the structure would indicate that a defensive strategy is the most appropriate.*
- *A risk analysis would justify this decision.*

4. Command Options

What are your considerations when selecting a Command option: Investigation, Fast Attack, or Command?

- *Due to the volume of fire and the need for an exterior attack, the IC should select the Command mode and direct resources to achieve the defensive strategy.*

5. Tactics

What are your considerations, if any, for the tactics (RECEO-VS) used to accomplish your incident priorities?

Rescue
- *The possibility that a savable victim is in the structure is reduced by the time of morning and the volume of fire.*
- *Firefighter life safety would take precedence justifying a defensive strategy.*
- *Therefore, search and rescue should not be attempted.*

Exposure
- *There are no exposures threatened by this fire.*
- *A spark patrol may be required if the embers start to spread due to the wind.*

Confinement
- *Master streams should be used to confine the fire to the building of origin.*

Extinguishment
- *Master stream appliances should be used to extinguish the majority of the fire.*
- *Handlines can be used to extinguish remaining fires.*
- *Class A or B foam may be used if available in the correct quantity to complete fire extinguishment.*

Overhaul
- *Overhaul will be applicable once the fire is extinguished.*

Ventilation
- *The structure has self-ventilated.*
- *Ventilation will not be required at this stage.*

Salvage
- *Salvage will be very limited due to the advanced stage of the fire.*

6. Unusual Hazards or Conditions

What are any unusual hazards or conditions which must be addressed? How will you address those hazards and/or conditions?

- *Unknown quantity and type of hazardous materials. Construction materials including paints, thinners, cleaners, and mastic may be present in unknown quantities.*
- *Natural gas lines may still be open, providing fuel under collapsed debris creating an explosion hazard.*
- *Lightweight wood truss construction. The structural integrity must be monitored by the incident commander and all firefighters working in or around the structure.*
- *Construction activities may have created openings in the floor slab or walls.*
- *Weakened roof and wall structures may collapse without warning.*
- *Unknown fire time that could lead to the potential for a building collapse.*
- *Time of response may contribute to responders not being completely alert.*

Scenario 5

On Sunday, January 24, at 3:00 p.m., dispatch reports a structure fire at the Super-Mart located at 1632 W. Frontage Rd. The dispatcher states that the fire started in the tire and lube section of the building. Current weather conditions are clear.

On Arrival
You are the first-arriving officer. You see a single-story, large area commercial building with black smoke showing from the Bravo side of the structure.

Water Supply
One blue 1,500 gpm (5 680 L/min) hydrant is located at the corner of N. Jardot and the North parking lot entrance.

One blue 1,500 gpm (5 680 L/min) hydrant is located at the corner of N. Jardot and the South parking lot entrance.

Weather
- Temperature – 52° F (11°C)
- Wind direction – South
- Wind velocity – 5 mph (8 km/h)
- Humidity – 71%

360° View

Alpha Side

Bravo Side

Charlie Side

Delta Side

Chapter 7 • Commercial Scenarios 289

Site Plan

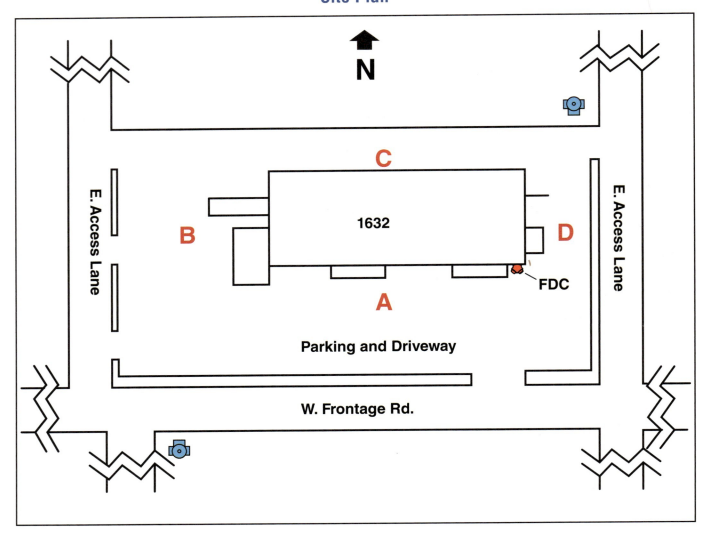

1. Size-Up

Based on the dispatch information and your visual observation of the scene on arrival, what are the facts and probabilities for this scenario?

Facts

- *Type II construction.*
- *Fire suppression (sprinkler) and standpipe system.*
- *Mercantile occupancy.*
- *Contents include combustible Class A as well as hazardous materials in all forms.*
- *Classified as a Target Hazard generating a preincident plan.*
- *Smoke is visible coming from the overhead doors of the vehicle maintenance area on the Charlie side of the structure.*
- *Water supply is close to structure with two 1,500 gpm (5 680 L/min) hydrants available.*
- *The structure is occupied and in the process of being evacuated.*

Probabilities

- *Fire suppression system may have been activated and be attempting to confine the fire. Smoke color and volume indicate that the system is not controlling the fire but may be confining it to the maintenance area.*
- *The area in which the fire is located contains flammable and combustible liquids, tires, compressed gases, plastics, and vehicles.*
- *The maintenance bay is separated from the remainder of the store by a barrier wall.*
- *Employees and customers may be trapped in or near the fire.*

2. Incident Priorities

What are your considerations for addressing the incident priorities of life safety, incident stabilization, and property conservation?

Life Safety

- *Search and rescue of any occupants located in or near the maintenance area are the first priority.*

Incident Stabilization

- *Rapid fire extinguishment is necessary to prevent the fire from extending into the sales area and the stock rooms.*
- *The fire suppression system should be supported and supplemented with attack hoselines.*

Property Conservation

- *Rapid fire extinguishment will prevent further damage to the structure or contents.*
- *Application of proper ventilation techniques will reduce smoke damage to the contents in the stock room and sales area.*
- *Use of salvage covers and moving some items will protect contents adjacent to the fire area from water and smoke damage.*

3. Operational Strategy

What are your considerations when selecting your operational strategy?

- *An offensive strategy is required to prevent loss of life and high content value loss.*

4. Command Options

What are your considerations when selecting a Command option: Investigation, Fast Attack, or Command?

- *Due to the size of the structure, the initial IC should select the Command mode and manage the incident from a Command Post.*

5. Tactics

What are your considerations, if any, for the tactics (RECEO-VS) used to accomplish your incident priorities?

Rescue
- *Search and rescue for victims near the fire should begin first.*
- *Store employees may be involved in the evacuation of customers but should not be assigned this task.*
- *Store manager can assist in accounting for employees.*

Exposure
- *The primary exposures are internal, consisting of the sales area and the stockrooms or other spaces adjacent to the maintenance area.*

Confinement
- *Confinement is initially performed by the fire suppression system. Resources must be assigned to support this system and to supply the standpipe system.*
- *Initial fire attack hoselines can be directed into the maintenance area through the overhead doors.*
- *Additional lines can be connected to standpipe outlets within the structure as resources become available.*

Extinguishment
- *A coordinated attack by hoselines from inside the structure and at the overhead door openings along with the fire suppression system should be mounted.*
- *Class A or B foam may be required to extinguish some burning materials.*

Overhaul
- *Removal of debris may be necessary to help locate hidden fires.*

Ventilation
- *Horizontal ventilation should be used to pull the smoke out of the maintenance area.*
- *Mechanical horizontal ventilation can be used from the barrier wall side.*
- *If necessary, vertical ventilation can be used over the fire.*
- *HVAC system needs to be secured to prevent it from spreading smoke and fire into the remainder of the structure.*

Salvage
- *Salvage covers and the removal of merchandise can be used to protect the contents in the sales area and stock room.*
- *Application of proper ventilation techniques can prevent smoke from entering the unburned areas.*
- *Retracted overhead doors must be monitored for structural stability.*

6. Unusual Hazards or Conditions

What are any unusual hazards or conditions which must be addressed? How will you address those hazards and/or conditions?

- *Potentially high occupant load.*
- *Exposed structural members.*
- *Hazardous materials.*
- *Potential for rapid fire spread into a large, uncompartmentalized area.*
- *Large open area.*
- *High fuel load.*
- *Vehicles present in the maintenance area.*
- *Overhead doors may be subject to collapse or closing.*

Notes

Notes

Scenario 6

On Saturday, October 15, at 7:00 p.m., dispatch reports a structure fire located at 2288 Cotton Tail Road. The dispatcher states that passersby called in the report and stated they saw heavy fire and smoke showing from the hayloft of a barn.

On Arrival
You are the first-arriving officer. You see a large barn structure with heavy smoke and fire showing from the hayloft of the structure.

Water Supply
No hydrant water supply in the immediate area

Pond is located behind structure

Weather
- Temperature – 58°F (°C)
- Wind direction – Northeast
- Wind velocity –10 mph (16 km/h)
- Humidity –75%

360° View

Alpha Side

Bravo Side

Charlie Side

Delta Side

Chapter 7 • Commercial Scenarios **297**

Site Plan

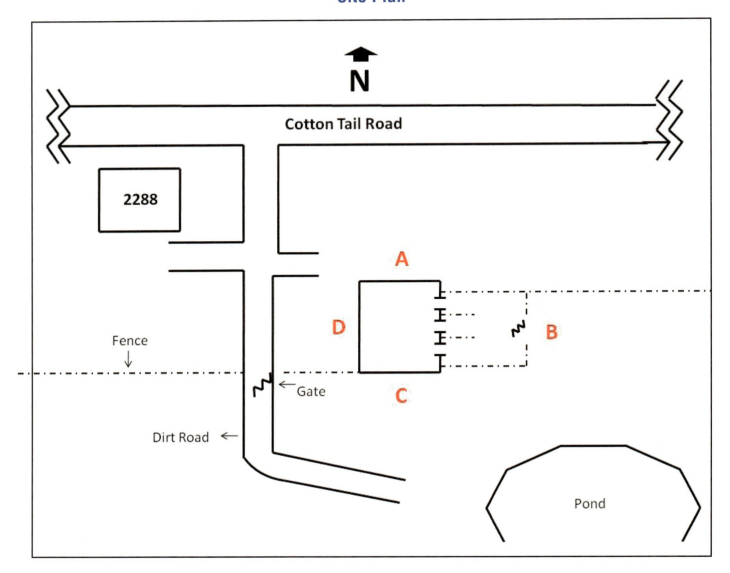

1. Size-Up

Based on the dispatch information and your visual observation of the scene on arrival, what are the facts and probabilities for this scenario?

Facts
- *Type V wood frame construction with metal siding and roof.*
- *Storage building containing hay and possibly livestock.*
- *Fire is visible from the loft door on the Alpha side.*
- *Water supply consists of a pond on the Charlie side of the structure.*
- *There are no exterior exposures.*
- *Electric lines provide power to the structure.*

Probabilities
- *Volume of fire in the hayloft will rapidly weaken both the roof rafters and joists and the floor structure.*

- *There is a potential for structural collapse.*
- *Initial assignment may not have enough water on hand to control the incident without establishing a shuttle operation or drafting from the pond.*
- *Establishing either a drafting operation or water shuttle will be time-consuming.*
- *Evacuation of livestock will require high level of staffing.*
- *Overhaul of structure will require a high level of staffing.*
- *Full fire extinguishment and overhaul will require copious amounts of water.*
- *Structural stability is unknown due to the fact that it may be wood or exposed steel.*
- *The barn may contain unknown quantities of farm chemicals including fertilizer, insecticides, and fuel.*

2. Incident Priorities

What are your considerations for addressing the incident priorities of life safety, incident stabilization, and property conservation?

Life Safety

- *A life safety hazard will exist in many forms. First is the safety of the firefighters. Interior operations will be very hazardous due to the location of the fire and the collapse potential of the hayloft. Second is the safety of any employees who may have entered the structure to remove livestock. Third is the safety of the livestock, reported to be horses. The monetary value of the latter is unknown.*
- *The volume and location of the fire would indicate the need for an offensive attack from the outside. However, if rescue of employees or evacuation of livestock is required, then fire attack hoselines will have to be extended into the structure through the opening on the Delta side.*

Incident Stabilization

- *If there is sufficient water supply available, rapid fire attack on the fire in the hayloft is required to limit the damage to the structure and decrease the potential for collapse.*
- *At the same time, the more water that is applied to the fire, the greater the weight on the hayloft floor, which will also increase the potential for collapse.*

Property Conservation

- *If the livestock is considered property rather than a life safety issue, the effort to evacuate them must be balanced against the risk to firefighters. Allowing employees to enter the structure will increase the life safety hazard.*
- *Removal of any farm equipment may not be possible due to the location of the fire over the main entrance.*
- *Rapid fire extinguishment should help to decrease the potential property damage.*
- *Exposure protection and preventing the fire from extending to the pasture downwind is a major concern. Containing a ground cover fire will require considerable water and resources.*

3. Operational Strategy

What are your considerations when selecting your operational strategy?

- *An initial offensive attack from the exterior on Alpha side will help to reduce the volume of fire.*
- *Fire crews must maintain situational awareness of conditions within the structure due to the volume of fire and the potential for collapse.*
- *There is a water supply close to the structure that should be adequate to support an offensive attack or a defensive strategy for exposure protection.*

4. Command Options

What are your considerations when selecting a Command option: Investigation, Fast Attack, or Command?

- *Depending on the resources available, the initial IC should operate in the Command mode and establish a strong Command presence at the Command Post.*

5. Tactics

What are your considerations, if any, for the tactics (RECEO-VS) used to accomplish your incident priorities?

Rescue

- *If rescue of any employees or evacuation of livestock is required, access should be gained through the openings on the Delta side.*
- *If rescue is required and the potential for collapse does not exist, a charged hoseline must be advanced to protect the rescue team.*
- *If rescue or evacuation is not required or the potential for collapse exists, crews should remain outside the structure.*

Exposure

- *The primary exposure concern is the internal unburned portion of the structure.*
- *Rapid fire extinguishment and the placement of hoselines at the openings on the Bravo and Delta sides can be used to limit the spread of fire into the rest of the structure.*
- *Exterior exposure is the dry grass surrounding the structure.*
- *Personnel with handlines or portable water extinguishers should be assigned for spark patrol as necessary.*

Confinement

- *Rapid attack and extinguishment should confine the fire to the loft area. However, the volume of fire and the potential for collapse may limit the ability to achieve this.*

Extinguishment

- *Direct application of water into the hayloft from the Alpha side should be used to extinguish the fire.*
- *The location and volume of the fire as well as the type of fuel on fire will require large (copious) amounts of water. This will increase the collapse hazard.*

Overhaul

- *To limit the collapse hazard, the hay must be removed from the loft as soon as it is extinguished. Prior to allowing firefighters into the barn on the ground floor or the loft, the structural integrity of the loft must be ensured.*

Ventilation

- *Ventilation is not an issue due to the type and configuration of the structure.*
- *Under no circumstances should vertical ventilation be attempted until the structural integrity of the roof is ensured.*

Salvage

- *If the livestock is classified as property, then evacuation may be considered a salvage tactic. Saddles, bridals, and other items stored in the barn can be removed if they are located in an area that will not pose a life safety risk to firefighters.*

6. Unusual Hazards or Conditions

What are any unusual hazards or conditions which must be addressed? How will you address those hazards and/or conditions?

- *Water supply challenges including time and staffing requirements*
- *Access to the structure*
- *Access to the static water supply in the pond*
- *Removal of livestock*
- *Heavy fuel load (hay)*
- *Location of fuel load*
- *Potential for collapse due to the amount of water required for extinguishment*
- *Overhaul challenges*
- *Delayed resources*
- *Limited or no additional resources*
- *Lack of appropriate apparatus to deal with limited water supply*
- *Potential for wildland fires caused by burning embers*
- *Potential collapse of hayloft caused by fire and age of structure*
- *Potential collapse of roof due to direct exposure to fire*
- *Potential for a variety of hazardous chemicals including fertilizer, insecticides, and fuel*

Special Hazards Scenarios

Chapter Contents

Scenario 1..............................313	Scenario 4..............................333
Scenario 2..............................319	Scenario 5..............................341
Scenario 3..............................327	Scenario 6..............................349

Divider page photo courtesy of Chief Chris Mickal, New Orleans (LA) Fire Department.

chapter 8

Scenarios

This chapter contains six hypothetical scenarios based on special occupancy types that are found in most response areas. The scenarios contain best practices that can be applied to each type of situation. However, these are suggestions only and not hard and fast rules that must be applied in every incident.

Because these scenarios are examples to help you learn how to make decisions based on the information in the previous chapters, there are limitations placed on them. The resources available to you will arrive within 10 minutes. You are the Incident Commander (IC) and must allocate your resources to control/mitigate the incident to the best of your ability. The resources you have reflect those in your local department, including first-alarm assignments, mutual aid, equipment, and staffing.

Special Hazards Scenarios

Learning Objectives

1. Given information on a special occupancy fire and resources available to your department, develop an Incident Action Plan (IAP) for achieving the incident objectives.

Note to Instructors

The following scenarios contain recommended practices based on the facts provided and on information contained in the text of this manual. These scenarios are intended to be used as teaching aids. It is suggested that you develop scenarios specific to your department's resources, procedures, and response area. Your scenarios should contain structures, hazards, water supplies, photographs, and site plans that exist in your jurisdiction. Do not include probabilities, incident priorities, strategies, command options, or tactics. This information should be supplied by the students based on their understanding of the text. Your scenarios can be varied based on changes to the water supply (hydrants out of service, peak water demand times), delayed resources, weather conditions, minimum staffing, time of day, or local standard operating procedures (SOPs).

Chapter 8
Special Hazards Scenarios

Scenario 1

On Saturday, June 19, at 3:00 p.m., dispatch reports a structure fire at the Hughes Lumber Yard located at 310 E. Third Street. Dispatch states that a call from the building manager reports that there is a fire in the employee break room located in the center of the building. All customers and employees have evacuated the building.

On Arrival
You are the first-arriving officer. You see a single-story commercial building with moderate smoke showing from the roof of the structure.

Water Supply
One red 500 gpm (1 900 L/min) hydrant at the corner of E. Third Street and Giordano Avenue

One red 500 gpm (1 900 L/min) hydrant at the corner of E. Third Street and Julius Avenue

Weather
- Temperature – 80°F (27°C)
- Wind direction – Northwest
- Wind velocity – 20 mph (32 km/h)
- Humidity – 75%

360° View

Alpha Side

Bravo Side

Charlie Side

Delta Side

Chapter 8 • Special Hazards Scenarios

Site Plan

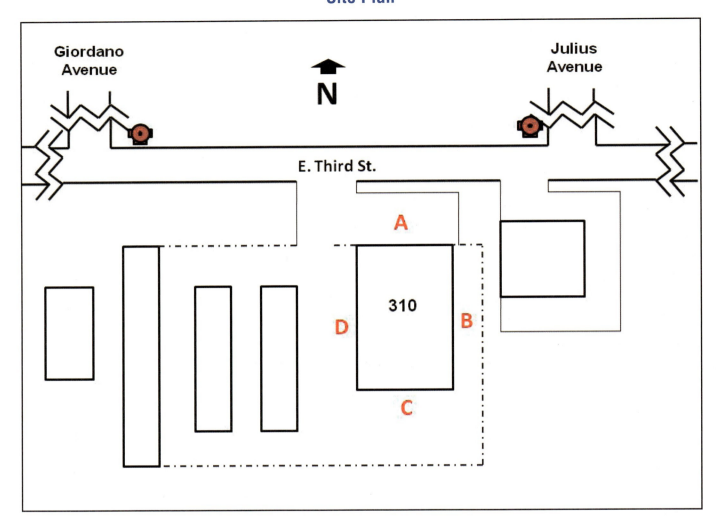

1. Size-Up
Based on the dispatch information and your visual observation of the scene on arrival, what are the facts and probabilities for this scenario?

Facts
- Type II construction.
- Mercantile occupancy.
- Classified as target hazard due to contents.
- All occupants have been evacuated.
- Smoke visible from the center section of the roof, possibly a vent.
- Wind is strong out of the northeast.
- Water supply is close to the structure with two 500 gpm (1 900 L/min) hydrants available.
- There are exposures on the Bravo and Delta sides.
- Contents may include combustible finishes, flammable liquids, compressed gas, and plastics.

Probabilities
- *Large building with an isolated area of smoke.*
- *Compartmentalized building with retail, wholesale, and storage.*
- *Fire may be confined to a small area.*
- *Fire appears to be in the early stage of development.*
- *Smoke may be exiting the roof through a roof vent, scuttle, or other opening as it is too early in the fire development for it to have breached the roof.*
- *Risk management evaluation; minimal risk with savable property.*
- *Contents can contribute to rapid fire spread.*

2. Incident Priorities

What are your considerations for addressing the incident priorities of life safety, incident stabilization, and property conservation?

Life Safety
- *Occupants have been evacuated.*
- *Life hazard to firefighters increases if fire is allowed to expand.*

Incident Stabilization
- *Rapid fire extinguishment is indicated by the location of the fire and the visible smoke.*

Property Conservation
- *Rapid fire extinguishment will reduce damage to the structure and contents.*
- *Proper application of ventilation techniques will decrease or limit smoke damage.*
- *Proper application of salvage techniques will decrease or limit smoke and water damage.*

3. Operational Strategy

What are your considerations when selecting your operational strategy?

- *The need for rapid fire extinguishment and evacuation requires an offensive strategy.*

4. Command Options

What are your considerations when selecting a Command option: Investigation, Fast Attack, or Command?

- *Depending on available resources, the initial IC should use the Fast Attack mode, supervising the attack hoseline team.*
- *If sufficient resources are available, the IC can assume the Command mode and direct operations from the Alpha side of the structure.*

5. Tactics

What are your considerations, if any, for the tactics (RECEO-VS) used to accomplish your incident priorities?

Rescue

- *Not required because occupants have been evacuated.*

Exposure

- *Due to the location and size of the fire, exterior exposures are not a concern at this time.*
- *The primary exposure concern is the showroom and storage area adjacent to the break room.*
- *This exposure can be protected through rapid fire extinguishment.*

Confinement

- *Rapid fire extinguishment can confine the fire to the room of origin.*

Extinguishment

- *Rapid fire extinguishment can confine the fire, protect the exposures, and extinguish the fire at the same time.*
- *Fire attack hoselines placed inside the structure and directed into the door of the break room are used for extinguishment.*

Overhaul

- *Removal of ceiling tiles and wall coverings in the break room is required to determine and eliminate fire extension.*

Ventilation

- *The fire appears to have ventilated through the roof, either through a vent or through an opening caused by the fire.*
- *Forcing air into the room will continue this process once the fire is extinguished.*

Salvage

- *Salvage covers should be placed over contents adjacent to the fire area.*

6. Unusual Hazards or Conditions

What are any unusual hazards or conditions which must be addressed? How will you address those hazards and/or conditions?

- *Roof designs and the use of lightweight trusses.*
- *Water supply may be a critical factor if the fire progresses based on only two 500 gpm (1 900 L/min) hydrants.*
- *Large area and open floor space may hide the extent of the fire involvement.*
- *Lack of fire protection systems.*
- *Heavy fire load with the potential for rapid fire spread.*
- *Unknown and various building contents.*
- *High ceilings.*

Scenario 2

On Saturday, October 31, at 3:00 a.m., dispatch reports a structure fire at a warehouse located at 2157 N. Jardot. Dispatch reports that the fire was reported by a passerby. The most recent preincident survey indicates that the building has been vacant for the last three months.

On Arrival
You are the first-arriving officer. You see a single-story commercial building with fire showing from the Alpha side of the structure.

Water Supply
One blue 1,500 gpm (5 680 L/min) hydrant at the corner of N. Jardot and the North parking lot entrance

One blue 1,500 gpm (5 680 L/min) hydrant at the corner of N. Jardot and the South parking lot entrance

Weather
- Temperature – 52°F (11°C)
- Wind direction – Northeast
- Wind velocity – 10 mph (16 km/h)
- Humidity – 68%

360° View

Alpha Side

Bravo Side

Charlie Side

Delta Side

Chapter 8 • Special Hazards Scenarios **313**

Site Plan

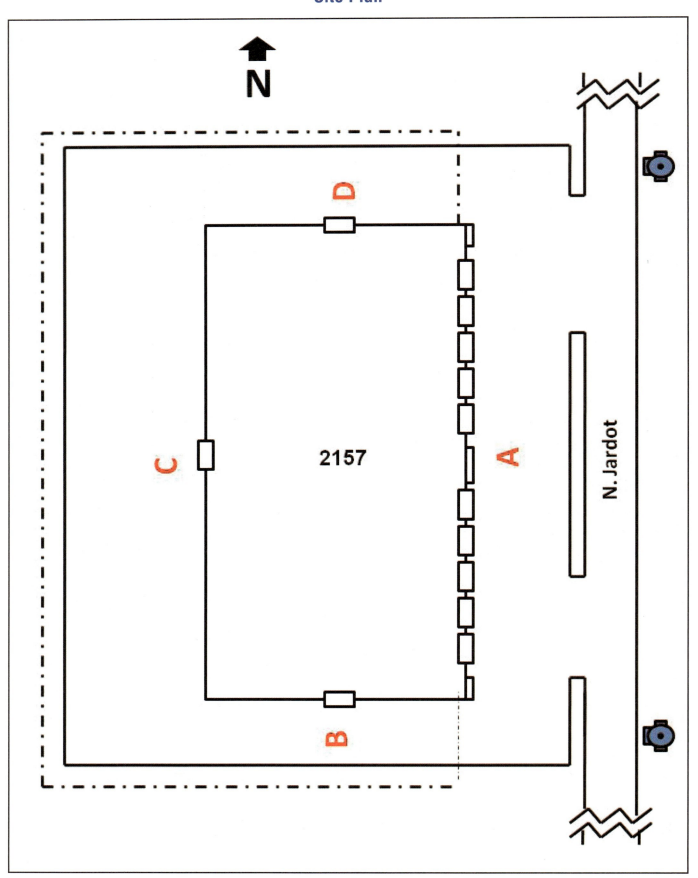

1. Size-Up

Based on the dispatch information and your visual observation of the scene on arrival, what are the facts and probabilities for this scenario?

Facts

- Type II construction.
- Exposed lightweight roof trusses.
- Warehouse structure.
- Fire is visible through an opening on the Alpha side near the Delta corner.
- Smoke is coming from a door on the Alpha Side and from the vent openings on the Charlie side.
- Smoke conditions indicate a fire in the growth stage.
- Wind is moderate out of the northeast.
- Water supply is close to structure with two 1,500 gpm (5 680 L/min) hydrants available.
- There are no exterior exposures.

Probabilities

- Smoke seems to be confined to one area on the Delta end of the structure, indicating the presence of barrier walls.
- Size and occupancy type would indicate the presence of a fire suppression system although no water flow alarm has been activated.
- Large area and open floor space may hide the extent of the fire involvement.
- Unknown type and quantity of contents.
- Possibility that the structure is occupied or that it has been illegally entered and vandalized.

2. Incident Priorities

What are your considerations for addressing the incident priorities of life safety, incident stabilization, and property conservation?

Life Safety

- Time of night and the type of occupancy would indicate that the structure is unoccupied.
- If there is evidence of forced or illegal entry, search and rescue may be initiated.

Incident Stabilization

- Rapid fire attack hoselines placed between the fire and the unburned portion of the structure.

Property Conservation

- Rapid fire attack will reduce or limit damage to the structure and contents.

3. Operational Strategy

What are your considerations when selecting your operational strategy?

- An offensive strategy is indicated due to the size of the fire and amount of visible smoke.

4. Command Options

What are your considerations when selecting a Command option: Investigation, Fast Attack, or Command?

- *Depending on the available resources, the Fast Attack Command option should be implemented.*
- *Once additional resources arrive, the initial IC should establish a Command Post and shift to the Command option.*

5. Tactics

What are your considerations, if any, for the tactics (RECEO-VS) used to accomplish your incident priorities?

Rescue

- *If there is evidence that the structure is occupied, search and rescue should be initiated when the fire is controlled.*
- *Hoselines should be used to assist search teams.*
- *Horizontal ventilation may be required to improve visibility within the structure.*
- *Depending on the internal configuration of the structure, additional search teams may be required.*

Exposure

- *There is no exterior exposure visible.*
- *The primary exposure is the unburned portion of the structure.*
- *Depending on the internal configuration of the structure, fire barrier walls may provide sufficient exposure protection.*
- *If no barrier walls exist, placing fire attack hoselines between the fire and the unburned portion of the warehouse can be used for exposure protection.*

Confinement

- *If the structure is equipped with an operating fire suppression system, it should be supported through the fire department connection to confine the fire to the area of origin.*
- *If a system does not exist or the system is inoperative, fire attack hoselines should then be used to confine the fire to the area of origin and cool the roof trusses.*

Extinguishment

- *A rapid fire attack at the seat of the fire is recommended for quick extinguishment.*
- *Attack hoselines may enter from the Delta side or any of the Alpha-side openings, depending on the interior configuration of the structure.*
- *Depending on the type of contents on fire, the use of Class A or B foam may be necessary.*

Overhaul
- *Removal of debris once the fire is extinguished may be required to prevent rekindle.*
- *Before any debris is removed, fire cause determination must be made.*

Ventilation
- *Horizontal ventilation must be coordinated with search and rescue teams and fire attack teams.*
- *If the roof is still structurally sound, vertical ventilation may be used through existing roof vents or opening cut in the roof.*

Salvage
- *Contents in unburned areas of the structure may require protection using salvage covers.*

6. Unusual Hazards or Conditions

What are any unusual hazards or conditions which must be addressed? How will you address those hazards and/or conditions?

- *Unknown quantity and type of contents.*
- *Potential quantities of hazardous materials.*
- *Lightweight building construction.*
- *Unknown existence or condition of fire protection system.*
- *High ceilings.*
- *Possible arson fire.*
- *Possibility of transient occupants.*
- *Possibility of illegal activity such as a meth lab.*

Notes

Notes

Scenario 3

On Wednesday, April 24, at 4:00 p.m., dispatch reports a structure fire at the Hillside Methodist church located at the corner of 4th Street and N. Stallard Road. The dispatcher states that the caller reported that the fire started in the mechanical room on the south side of the building and that no occupants are in the structure. Current weather is clear.

On Arrival

You are the first-arriving officer. You see a single-story church building with fire and smoke showing from the Charlie side of the structure.

Water Supply

One blue 1,500 gpm (5 680 L/min) hydrant is located at the corner of N. Stallard and 4th Street.

Weather

- Temperature – 65°F (18°C)
- Wind direction – North
- Wind velocity – 5 mph (8 km/h)
- Humidity – 62%

360° View

Alpha Side

Bravo Side

Charlie Side

Delta Side

Chapter 8 • Special Hazards Scenarios

Site Plan

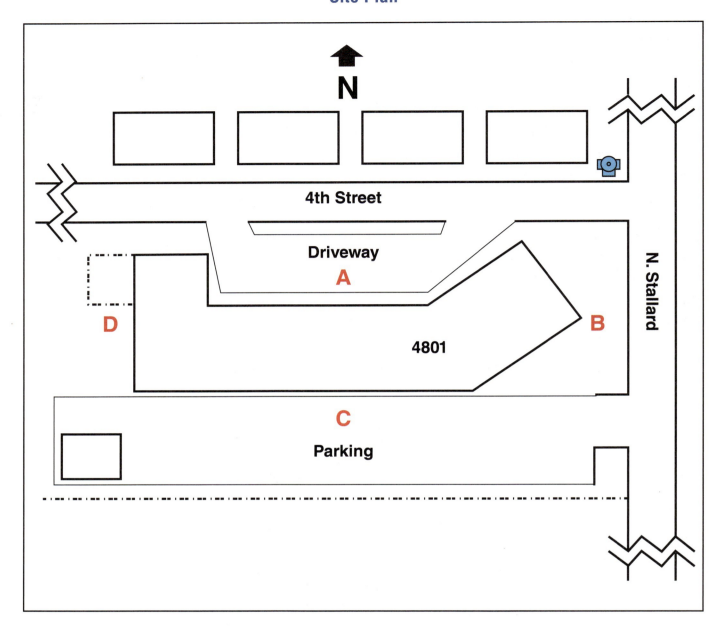

1. Size-Up

Based on the dispatch information and your visual observation of the scene on arrival, what are the facts and probabilities for this scenario?

Facts

- *Type III construction.*
- *Lightweight roof truss construction.*
- *Structure is an Assembly classification.*
- *Occupants have self-evacuated.*
- *Fire is located in a mechanical room near the back of the structure.*
- *Sanctuary is a large open area with a high ceiling remote from the scene of the fire.*

- *Smoke is coming from a window on Charlie side.*
- *Water supply is close to structure with one 1,500 gpm (5 680 L/min) hydrant available.*
- *There is an exposure located across the parking lot on the Charlie side.*

Probabilities

- *Fire may be confined to the mechanical space.*
- *Roof trusses may be affected by the fire creating a collapse hazard.*
- *Fire may spread through the cockloft to other areas.*
- *Interior finishes may contribute to fire spread.*
- *Potential for additional fuel from heating equipment, such as natural gas.*

2. Incident Priorities

What are your considerations for addressing the incident priorities of life safety, incident stabilization, and property conservation?

Life Safety

- *Search and rescue are not indicated since occupants have self-evacuated.*

Incident Stabilization

- *Rapid fire attack is indicated as a means to limit fire development and spread to the mechanical room.*

Property Conservation

- *Rapid fire attack is indicated as a means to limit fire development and spread to the mechanical room.*
- *Rapid extinguishment will reduce damage to contents and structure.*

3. Operational Strategy

What are your considerations when selecting your operational strategy?

- *Your choice of strategy may depend on how many resources you will have quickly available at the scene. There is a direct route into the fire area through the front door which, would make an offensive operation straightforward. If an offensive strategy is selected, adequate resources must be available to quickly confine and extinguish the fire as well as addressing any fire extension to the attic.*
- *Another consideration is the exposure on the Delta side.*
- *A water supply is close to the structure that should be adequate to support an offensive attack or a defensive strategy for exposure protection.*

4. Command Options

What are your considerations when selecting a Command option: Investigation, Fast Attack, or Command?

- *Depending on the availability of resources, the initial IC may select the Fast Attack option.*
- *As additional resources arrive, the IC should transition to the Command option in order to direct the additional personnel and units.*

5. Tactics

What are your considerations, if any, for the tactics (RECEO-VS) used to accomplish your incident priorities?

Rescue
- *Search and rescue are not indicated due to the self-evacuation of the occupants.*

Exposure
- *The only external exposure is located away from the fire area and is not initially a concern.*
- *The internal exposures are other work areas, kitchen, library, meeting rooms, and the sanctuary.*
- *Rapid fire attack from the interior side of the fire can be used to protect these exposures.*

Confinement
- *Rapid fire attack from the interior of the structure can confine the fire to the mechanical room.*

Extinguishment
- *Rapid fire attack from the interior of the structure to extinguish the fire in the mechanical room.*

Overhaul
- *Ceiling tiles will need to be removed to determine fire spread in attic and damage to the roof trusses.*
- *Walls adjacent to the fire should be opened to locate any hidden fires.*

Ventilation
- *Horizontal ventilation must be coordinated with the attack hose teams.*
- *Interior doors should be closed to prevent smoke spread.*
- *HVAC should be shut off to limit smoke spread.*

Salvage
- *Interior doors should be shut to limit smoke spread.*
- *Once fire cause has been determined, debris should be removed from the fire area.*
- *Salvage covers and floor runners can be used to protect contents in adjacent rooms from water and smoke damage.*

6. Unusual Hazards or Conditions

What are any unusual hazards or conditions which must be addressed? How will you address those hazards and/or conditions?

- *Large open area of sanctuary may contribute to increase in fire intensity.*
- *Lightweight roof trusses over mechanical room may weaken and collapse.*
- *Interior finishes and hidden spaces above ceilings can contribute to fire spread.*

Scenario 4

On Friday, August 15, at 6:00 p.m., dispatch reports a structure fire at The Old Towne Bar and Grill located at 208 W. Walnut. The dispatcher states the caller reported that the fire started in the kitchen. Current weather condition is clear.

On Arrival
You are the first-arriving officer. You see a two-story assembly occupancy building with fire and smoke showing from the Bravo side of the structure.

Water Supply
One blue 1,500 gpm (5 680 L/min) hydrant is located one block North on W. Walnut

Weather
- Temperature – 85° F (29°C)
- Wind direction – South
- Wind velocity – 5 mph (8 km/h)
- Humidity – 75%

360° View

Alpha Side

Bravo Side

Charlie Side

Delta Side

Chapter 8 • Special Hazards Scenarios

Site Plan

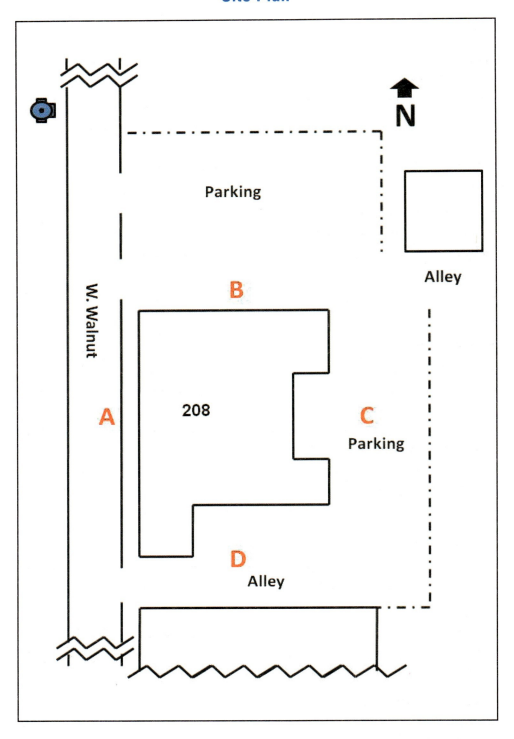

1. Size-Up

Based on the dispatch information and your visual observation of the scene on arrival, what are the facts and probabilities for this scenario?

Facts

- Type III construction.
- Assembly occupancy.
- Structure consists of multiple buildings attached to each other with additions to the Bravo and Charlie sides.
- Structure has been remodeled many times over its lifetime.
- Due to time of day and day of the week, the nightclub is near capacity.
- Fire is reported to be located in the kitchen.
- Smoke is coming from the Bravo side of the structure.
- Water supply is close to structure with one 1,500 gpm (5 680 L/min) hydrant available.
- There is an exposure on Delta side.

Probabilities

- Fire may involve cooking oils or grease.
- Employees may be in the kitchen area attempting to control the fire.
- Due to the time of day and day of the week, customers may need assistance in evacuation.
- Fire may spread rapidly into other areas of the structure through voids or penetrations of walls or ceilings.
- Interior configuration may consist of narrow or dead-end corridors, multiple-floor levels, as well as flammable finishes.
- Customers may be unfamiliar with the configuration or exit paths.

2. Incident Priorities

What are your considerations for addressing the incident priorities of life safety, incident stabilization, and property conservation?

Life Safety

- Evacuation of customers and employees is the primary priority.
- Firefighter safety is required due to the age and internal configuration of the structure.

Incident Stabilization

- Rapid fire extinguishment is required to protect the search and rescue team and prevent fire spread.

Property Conservation

- Rapid fire extinguishment will limit fire spread and smoke damage.

3. Operational Strategy

What are your considerations when selecting your operational strategy?

- *Due to the high life safety factor, an offensive strategy is required. Customers are unfamiliar with the structure, may be impaired, and will tend to use the same entrance as their emergency exit.*

4. Command Options

What are your considerations when selecting a Command option: Investigation, Fast Attack, or Command?

- *Due to the complexity of the structure and the high occupant load, the initial IC should adopt the Command option to effectively direct resources in search and rescue and fire extinguishment.*

5. Tactics

What are your considerations, if any, for the tactics (RECEO-VS) used to accomplish your incident priorities?

Rescue

- *The primary tactic is to evacuate the customers and employees from the structure. The majority of the customers will attempt to use the front entrance to leave the building.*
- *Employees may be assisting with evacuating and directing customers to alternate exits upon arrival.*
- *Employees may be required to open or clear exit paths if necessary.*

Exposure

- *The initial exposures are internal to the structure. First the areas adjacent to the kitchen, then the other attached buildings.*
- *Placement of an attack fire hose between the adjacent areas where occupants and employees may still be located and the kitchen will protect these individuals as well as protect egress routes.*

Confinement

- *Attack fire hoselines placed on the interior of the structure can help to confine the fire to the kitchen.*

Extinguishment

- *Attack fire hoselines placed on the interior and directed into the kitchen can be used for fire extinguishment.*
- *If it is difficult to advance a hoseline into the structure from the Alpha side, an attack could be made directly from the Charlie side into the kitchen. Care must be taken to keep from spreading the fire into the rest of the building.*

Overhaul

- *Ceiling tiles above the attack hoselines on the interior of the structure should be removed to monitor fire extension in the cockloft.*
- *Ceilings and walls must be opened to locate fire extension into hidden areas of the structure.*
- *Overhaul must be rapid to prevent extension.*

Ventilation
- Horizontal ventilation should be used to pull the smoke from the kitchen on the Charlie side.

Salvage
- Salvage covers should be placed over contents in the remainder of the structure to protect from smoke and water damage.

6. Unusual Hazards or Conditions

What are any unusual hazards or conditions which must be addressed? How will you address those hazards and/or conditions?

- An EMS/Ambulance task force may be needed for the occupancy load.
- Gas-fueled kitchen equipment.
- Building construction and /or alterations with a lack of fire protection.
- High probability of impaired occupants.
- Crowded or complex interior arrangements will make it difficult to maneuver.
- High occupancy load.
- High volume interior arrangement.
- Heavy fuel load.
- Confusing interior arrangement which may affect self-extrication of patrons.
- Building construction will include multiple ceilings, void spaces, small interior stairwells, breached fire walls, and exposed or unprotected building components.
- Multiple avenues for fire travel and extension.
- Parapet wall creates collapse potential if weakened.
- Occupants are unfamiliar with the occupancy.

Notes

Notes

Scenario 5

On Tuesday, May 15, at 7:00 p.m., dispatch reports a fire alarm at the Anderson Inn located at 4801 S. Anderson Road. Current weather conditions are clear.

On Arrival

You are the first-arriving officer. You see a four-story hotel structure with nothing showing.

Water Supply

One blue 1,500 gpm (5 680 L/min) hydrant is located on the northeast side of the parking lot.

One blue 1,500 gpm (5 680 L/min) hydrant is located on the s outheast side of the parking lot.

Weather

- Temperature – 75° F (24°C)
- Wind direction – South
- Wind velocity – 5 mph (8 km/h)
- Humidity – 75%

360° View

Alpha Side

Bravo Side

Charlie Side

Delta Side

Site Plan

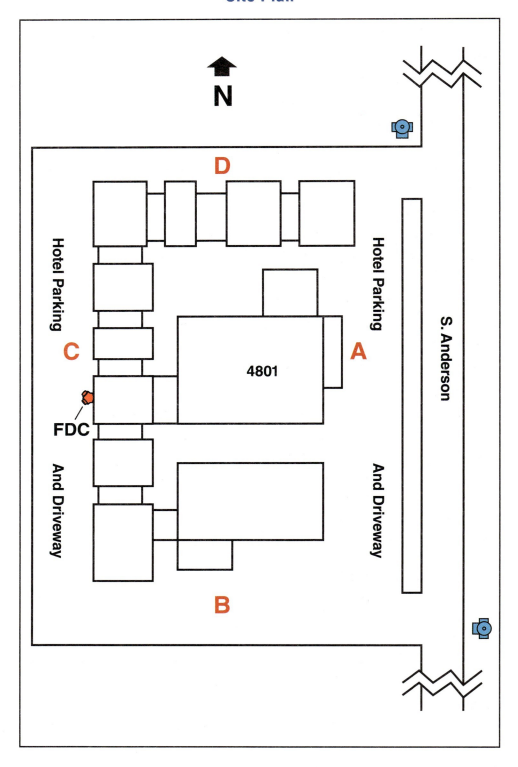

1. Size-Up
Based on the dispatch information and your visual observation of the scene on arrival, what are the facts and probabilities for this scenario?

Facts
- *Type II construction.*
- *Residential multifamily occupancy.*
- *Structure is occupied.*
- *Structure is new construction and built to code.*
- *Structure is required by code to have a fire suppression system and standpipe system supplied by a fire department connection (FDC).*
- *Water flow alarm has not been activated.*
- *Structure is required to have a fire control panel.*
- *Occupants are generally unfamiliar with the structure.*
- *A small number of employees are on-site.*
- *There are no visible indicators of smoke or fire.*
- *Water supply is close to structure with two 1,500 gpm (5 680 L/min) hydrants available.*
- *There is an exposure on Delta side.*

Probabilities
- *Occupants may not be familiar with the structure or exit paths.*
- *Exit and emergency lighting should be operational.*
- *Fire suppression system should be operational.*
- *Although the local fire alarm should have activated, occupants may not have evacuated the structure.*
- *A high number of personnel will be required to search and evacuate occupants and locate the source of the alarm.*

2. Incident Priorities
What are your considerations for addressing the incident priorities of life safety, incident stabilization, and property conservation?

Life Safety
- *The high occupant load and the potential for occupants who are unfamiliar with the egress paths increases the life safety hazard.*

Incident Stabilization
- *The initial action should be to determine if a fire hazard exists and where it is located.*
- *The fire control panel will help to locate the activated smoke detector.*
- *The location may indicate the need to place a hoseline into service from the nearest standpipe in the event that a fire exists.*

Property Conservation
- *Rapid fire location and extinguishment will limit any damage to the structure and content.*

3. Operational Strategy

What are your considerations when selecting your operational strategy?

- *Due to the lack of information on the location and extent of the hazard, an offensive strategy is required.*
- *Search for the fire combined with evacuation must be implemented promptly.*

4. Command Options

What are your considerations when selecting a Command option: Investigation, Fast Attack, or Command?

- *The initial IC should implement the Investigation mode by going to the fire control panel and then proceeding to the location indicated.*
- *This action may place the IC in the Fast Attack mode if a fire is discovered.*
- *If there are sufficient resources to assign to Investigation, the IC may establish Command at the fire control panel and direct search, rescue, and fire extinguishment from that location.*

5. Tactics

What are your considerations, if any, for the tactics (RECEO-VS) used to accomplish your incident priorities?

Rescue
- *Evacuation of the structure occurs simultaneously with the search for the source of the alarm.*
- *If the structure is equipped with a voice notification system, occupants should be directed to leave the building. They may also be directed to remain in place if they are not on the floor in which the alarm is activated.*

Exposure
- *All exposures are internal.*
- *Until the fire is located and confirmed, exposure activities are not implemented.*

Confinement
- *Until the fire is located and confirmed, confinement activities are not implemented.*

Extinguishment
- *Until the fire is located and confirmed, extinguishment activities are not implemented.*
- *If a fire is confirmed, fire attack hoselines must be deployed from the nearest standpipe connection in accordance with the local SOP/SOG.*
- *Local SOP/SOG will define responsibility for supplying the FDC depending on resources available.*

Overhaul
- *Until the fire is located and confirmed, overhaul activities are not implemented.*

Ventilation
- *Until the fire is located and confirmed, ventilation activities are not implemented.*

Salvage
- *Until the fire is located and confirmed, salvage activities are not implemented.*

6. Unusual Hazards or Conditions

What are any unusual hazards or conditions which must be addressed? How will you address those hazards and/or conditions?

- *High occupant load poses great life hazard.*
- *Size and configuration of the structure will increase evacuation time.*
- *Size and configuration of the structure will increase time required to locate the source of the alarm.*

Notes

Notes

Scenario 6

On Tuesday, December 18, at 10:00 p.m., dispatch reports a structure fire at the Fontana Inn hotel located at 412 W. Frontage Road. The dispatcher stated that the night manager called in the report. The night manager reported that the fire is located in a room on the fourth floor. The fourth floor corridor is filling with smoke, and the building's fire suppression system has apparently not activated. Current weather condition is snowing.

On Arrival
You are the first-arriving officer. You see a four-story hotel structure with fire showing from the Alpha/Bravo corner sides of the structure.

Water Supply
One blue 1,500 gpm (5 680 L/min) hydrant is located at the corner of S. Murphy and W. Frontage Road.

One blue 1,500 gpm (5 680 L/min) hydrant is located on the North East side of the hotel parking lot.

Weather
- Temperature – 28° F (2°C)
- Wind direction – North
- Wind velocity – 5 mph (8 km/h)
- Humidity – 100%

360° View

Alpha Side

Bravo Side

Charlie Side

Delta Side

Site Plan

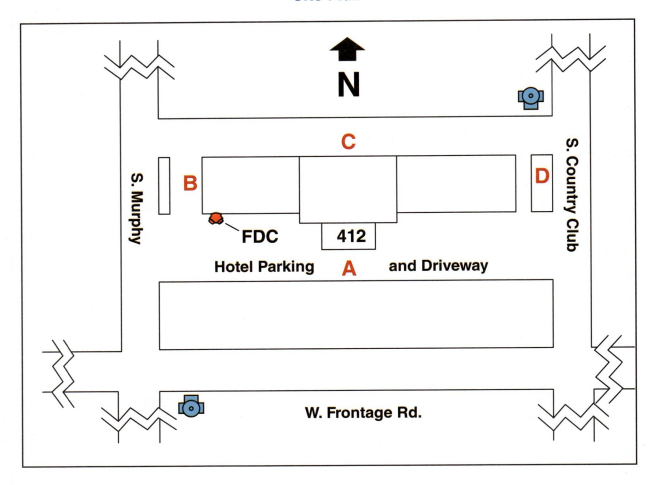

1. Size-Up

Based on the dispatch information and your visual observation of the scene on arrival, what are the facts and probabilities for this scenario?

Facts

- *Type II construction.*
- *Residential multifamily occupancy.*
- *Structure is occupied.*
- *Structure is new construction and built to code.*
- *Structure is required by code to have a fire suppression system and standpipe system.*
- *Water flow alarm has not been activated.*
- *Structure is required to have a fire control panel.*
- *Access doors are located on all four sides with stairwells on Bravo and Delta sides.*
- *Fire is visible in a room on the Alpha/Bravo corner of the fourth floor.*
- *No smoke is visible on the exterior of the structure.*

- *Water supply is close to structure with two 1,500 gpm (5 680 L/min) hydrants available.*
- *There is an exposure on Delta side.*

Probabilities
- *Time of night may indicate that some occupants may be asleep.*
- *Limited staffing by hotel employees.*
- *Fire suppression system may be inoperative.*
- *Occupants may not have been alerted to the hazard.*
- *The exit doors on the Bravo and Delta sides may be locked from the inside.*

2. Incident Priorities
What are your considerations for addressing the incident priorities of life safety, incident stabilization, and property conservation?

Life Safety
- *Due to the high occupant load and time of night, life safety is the primary priority.*

Incident Stabilization
- *Rapid fire extinguishment is required to protect occupants.*

Property Conservation
- *Rapid fire extinguishment is required to limit property damage.*

3. Operational Strategy
What are your considerations when selecting your operational strategy?
- *The need for search, rescue, and evacuation and fire extinguishment requires an offensive strategy.*

4. Command Options
What are your considerations when selecting a Command option: Investigation, Fast Attack, or Command?
- *Due to the high life safety hazard, the size of the structure, and the lack of confirmed fire protection, the initial IC should use the Command option and manage the incident from a Command Post or from the fire control panel in the lobby.*

5. Tactics
What are your considerations, if any, for the tactics (RECEO-VS) used to accomplish your incident priorities?

Rescue
- *Evacuation of the structure occurs simultaneously with fire extinguishment.*
- *If the structure is equipped with a voice notification system, occupants should be directed to leave the building. They may also be directed to remain in place if they are not on the floor in which the alarm is activated.*
- *Elevators must be controlled to prevent occupants from using them.*

- *A significant life safety threat exists. It appears the fire protection systems are not operating due to amount of fire and no water flow alarm. The best strategy is to shelter in place the fourth floor occupants, make very rapid fire attack to protect stairwell (first), hallway on fourth (second). The occupants on the other floors need to evacuate the building in case fire cannot be contained.*

Exposure
- *There are no apparent external exposures.*
- *Interior exposures include the rooms and exit hallway adjacent to the fire room and the roof immediately above.*

Confinement
- *Attack fire hoselines should be deployed per local SOP/SOG from the nearest standpipe.*

Extinguishment
- *Attack fire hoselines should be deployed per local SOP/SOG from the nearest standpipe.*

Overhaul
- *Ceiling tiles or sheetrock in the hallway and room adjacent to the fire should be removed to determine if the fire has extended in those directions.*
- *Ceiling tiles or Sheetrock™ and walls in the fire room should be opened once the fire is under control.*

Ventilation
- *The structure's ventilation system may need to be used to remove smoke from the structure or to pressurize other floors and stairwells.*
- *If the stairwells are equipped with ventilation fans, these can also be used.*
- *As a last resort, the window on the fire room can be removed for horizontal ventilation.*

Salvage
- *Salvage covers and floor runners should be placed over contents in the room adjacent to the fire and on the floor below the fire.*

6. Unusual Hazards or Conditions
What are any unusual hazards or conditions which must be addressed? How will you address those hazards and/or conditions?

- *High occupant load poses great life hazard.*
- *An EMS/Ambulance task force may be needed for the occupancy load.*
- *Coordination with a responsible party from the facility.*
- *Fire protection system does not appear to be in service.*
- *Weather considerations create slip hazards.*
- *May need additional resources.*
- *Stairwell considerations.*
- *Elevator control will need to be accomplished immediately.*

- *High probability of occupants unfamiliar with the occupancy and routes of exit.*
- *Occupants may be sleep impaired.*
- *Ventilation may not be possible if windows do not open or system is not designed to remove smoke.*

Notes

Appendices

Contents

Appendix A
　　Preincident Planning Checklist 351

Appendix B
　　Sample SOP/SOG for Engine Company
　　　Operations 354

Appendix C
　　NIOSH: Death in the Line of Duty Report 366

Appendix D
　　Tactical Worksheets 383

Appendix A
Preincident Planning Checklist

Cleveland Volunteer Fire Department

Pre-Incident Planning Checklist

Date of Inspection:	Committee Officer:
Committee Members:	

General Information
Facility/ Business Name:	
Street Address:	Nearest Cross Street:

Contact Information
Facility Phone Number: ( )	Other Phone Number: ( )
Business Owner:	Location:
Phone Number: ( )	Primarily works on site: yes ☐ no ☐
Mobile Number: ( )	Other:

Emergency Contacts
Name:	Title:
Location:	Phone Number: ( )
Mobile Number: ( )	Other:

Name:	Title:
Location:	Phone Number: ( )
Mobile Number: ( )	Other:

Name:	Title:
Location:	Phone Number: ( )
Mobile Number: ( )	Other:

- *If more room is required for emergency contacts, please use the back of this form.*

GPS Information
Latitude:	Longitude:
Method of determining latitude and longitude:	
Description of location identified by latitude and longitude:	

Construction Information

SIZE		STORIES		BUILDING STATUS	
Length:		Above Ground:		Under Construction: ☐	Vacant & Secured: ☐
Width:		Below Ground:		Occupied: ☐	Vacant & Unsecured: ☐
Area:				Idle (Not Routinely Used): ☐	

Cleveland Volunteer Fire Department - 103 S. Pearman Ave. - PO Box 602 - Cleveland, MS 38732 – (662) 843-3159 - www.clevelandfd.com

Cleveland Volunteer Fire Department

CONSTRUCTION TYPE			
Fire Resistive: ☐	Unprotected Non-Combustible: ☐	Protected Ordinary: ☐	Protected Wood Frame: ☐
Heavy Timber: ☐	Protected Non-Combustible: ☐	Unprotected Ordinary: ☐	Unprotected Wood Frame: ☐
Walls: _____	**Floors:** _____		**Roof:** _____

ROOF COVERING	
Tile (clay, cement, slate, etc.): ☐	Wood Shingles (treated / untreated): ☐
Composite Shingle (asphalt): ☐	Built Up: ☐
Metal: ☐	No Roof: ☐

OTHER CONSTRUCTION INFORMATION	
Exterior Features:	Date of last known modification:
Architect:	Facility built date:
Construction Company:	Supplied site plans: yes ☐ no ☐

Hazardous Material Specific Information	
Tier II Facility: yes ☐ no ☐	Up to date Chemical Inventory List: yes ☐ no ☐
MSDS received with Tier II forms: yes ☐ no ☐	Chemical Inventory List provided: yes ☐ no ☐

HAZARDOUS MATERIAL STORAGE			
Chemical Name	ID#	Quantity	Location

Utility Services Information	
Electric Meter Location:	
Natural Gas Meter Location:	
Breaker Panel Location(s):	
Heated by:	Location:
Water Heater type: Natural Gas ☐ Electric ☐	Location:

Cleveland Volunteer Fire Department - 103 S. Pearman Ave. - PO Box 602 - Cleveland, MS 38732 – (662) 843-3159 - www.clevelandfd.com

Cleveland Volunteer Fire Department

Alarm System

Alarm Present: yes ☐ no ☐	Automatic: ☐	Manual Pull Station: ☐	Combination: ☐

DETECTOR TYPE		POWER SUPPLY	
Smoke: ☐	Heat: ☐	Battery: ☐	Hardwire w/ Battery Backup: ☐
Carbon Monoxide: ☐	Combination: ☐	Plug In: ☐	Plug In w/ Battery Backup: ☐
Sprinkler w/ Water - Flow Detection: ☐		Hardwire: ☐	

Alarm Panel Location:
Alarm Company: Phone Number:

Water Supply Information

Sprinkler Riser: yes ☐ no ☐   Location:
Sprinkler Standpipe Connection: yes ☐ no ☐ Location:

SYSTEM TYPE

Wet Pipe: ☐	Dry Chemical System: ☐	Halogen System: ☐	Class K System: ☐
Dry Pipe: ☐	Foam System: ☐	CO2 System: ☐	Standpipes: ☐

Hydrant Location(s):
Hydrant Flow Rate(s):

Red (500gpm or less) ☐	Orange (500gpm to 1000gpm) ☐	Green (1000gpm to 1500gpm) ☐	Blue (1500gpm or greater) ☐

Special Hazards

Special Notes

- *If more room is required for notes, please use the back of this form.*

Cleveland Volunteer Fire Department - 103 S. Pearman Ave. - PO Box 602 - Cleveland, MS 38732 – (662) 843-3159 - www.clevelandfd.com

Appendix B
Sample SOP/SOG for Engine Company Operations

Courtesy of Fairfax County (VA) Fire and Rescue

1	**INTRODUCTION**
1.1	Background
1.1.1	Throughout the nineteenth and early parts of the twentieth century, firefighters controlled and extinguished fires by projecting streams onto a fire from the exterior of a structure. As a result, firefighters would not be able to meet their objective until either unreachable areas burned themselves out, or collapse occurred, thus exposing the fire to outside streams. This often resulted in exposure protection as the primary objective. Major improvements have been made in the last half of the twentieth century. The use of breathing apparatus, ventilation techniques, and protective clothing has allowed firefighters to operate on the interior of structure fires. **It should be remembered that the <u>primary function</u> of today's engine companies is to obtain and deliver a sufficient quantity of water in the safest and most effective configuration to the fire area to extinguish the fire**.
1.2	Purposes:

- To provide guidelines and general information regarding engine company operations
- To describe the duties and responsibilities of the engine company
- To identify tactical and strategic considerations for engine company operations
- To establish guidelines for fire stream positioning
- To define the engine company officer's roles and responsibilities
- To establish guidelines for apparatus positioning on the fireground
- To establish procedures for engine driver/operator qualifications

2	**PLANNING AND PREPARATION**
2.1	Planning and preparation starts the moment members assigned to the engine company arrive at work. Members should check for any pertinent information such as street closings, hydrants out-of-service, water main repairs, and standpipe and sprinkler systems that are out-of-service.
2.2	All tools and equipment not specifically assigned to a particular firefighter shall be checked to ensure that they are present and functioning properly. Additionally, firefighters shall ensure that the nozzles are inspected for proper settings and are functioning properly.
3	**RESPONDING**
3.1	Typical Response and Arrival Considerations
3.1.1	Whether an engine company is responding alone or as part of a larger assignment determines many of the actions, enroute preparations, and thought processes that take place during the response. Officers must be cognizant of assignments identified in specific operating books.
3.1.2	Engine companies must allow the Truck access to the front of the building to the truck companies. Room should be allowed for ladder companies to deploy their tools and ladders. Engine

company officers shall communicate the water supply locations along with the type and method of hose layout. Additionally, any other pertinent information that may affect the operations on the fireground must be communicated to other units that are enroute or on the scene.

3.2 Single Engine Response

3.2.1 In addition to all of the preparations generic to any call, members must be prepared to confront, for a variety of reasons, an incident that requires more than a single engine. This type of situation might then require the engine company crew to perform tasks not normally associated with engine company work, i.e. ventilation to facilitate rescue, ground ladder placement, etc.

3.2.2 In situations where an engine is operating by itself for some time, alternative hose techniques should be considered. As an example, one option would be to drop off a sufficient amount of hose or standpipe pack at the fire, connect the supply line to that hose, and have the engine reverse lay to the water source.

3.3 Multi-Engine Response

3.3.1 The engine company's assignment is dependent upon where it is in the dispatch order for the alarm. In other words, the responsibility of the company is based upon it being first due, second due, etc. This assignment is paramount in determining the preparations to be made by the engine company crew while enroute to the incident. In most instances, the company assignments will fall in the dispatch order. If a unit is out of position or other circumstances indicate it will arrive on the scene significantly before or after it normally would, the unit shall communicate via radio with the controlling dispatch center and advise of the change in assignments.

3.3.2 First-Due Engine

- Engine companies shall assess primary water supply needs and identify the source
- Current conditions should be assessed and reported to other units on the alarm. Give a preliminary "on-scene" radio report consisting of type of structure and evident conditions and a command statement
- Determine an initial strategy, mode of operation, and any additional resources needed
- Determine the tactics necessary to carry out the strategy after viewing as many sides of the structure as possible
- As a crew, take the actions necessary to implement the tactics. Communicate these actions in your "situation report" so all responding units are aware of your location and intentions

Example: "Engine 209 is on the scene of a two-story colonial with fire out of two windows, second floor, side Adam. We will be in an offensive mode of operation stretching a 1 ¾ through the front door to floor 2 with a crew of three. Need the truck to ladder the second floor on sides Adam and Charlie and commence primary search."

3.3.3 Second-Due Engine

- Closely monitor all radio traffic, particularly that of the first arriving unit(s)
- Provide water supply to other units or fire suppression systems
- Be prepared to assume command where appropriate
- Prepare to stretch a back-up line or to assist the first engine with their hose deployment

3.3.4 Third-Due Engine

- Position at the rear (Side Charlie) of the structure where appropriate
- Lay supply lines from an alternate water source to the rear of the structure
- Monitor all radio traffic
- Continuously monitor and assess needs

- Keep the incident commander apprised of situation in the rear of the structure

3.3.5 Fourth-Due Engine
- Supply water to third engine if needed
- Report to command for any assignments
- Assume role of R.I.T., except on high-rise fires

3.3.6 First-Due Engine, Second Alarm
- In most cases, the first-due engine on the second alarm shall establish staging unless the unit has received different orders from command. The officer shall select and announce the location of staging unless the location was determined prior to their arrival. The driver of this unit should obtain the staging officer vest and assume that role. <u>Note that in high-rise, mall, and some other special fire situations, this function will be known as "base" since "staging" is located within the building</u>
- The officer and remaining crewmembers shall report to command for assignment, unless needed to manage staging
- It is recognized that many times greater alarm units are given tactical assignments before arriving at staging. In these cases, the responsibilities outlined above will fall to the first engine company that actually arrives in staging.

3.4 Monitoring Radio Traffic

3.4.1 Companies responding to alarms shall monitor radio traffic from the dispatch center as well as units that are operating on-scene. This will provide firefighters with vital information about conditions at the scene and make them aware of the problems encountered by first arriving units.

3.5 Safety

3.5.1 All responses must be made consistent with all current Commonwealth of Virginia driving laws as well as departmental regulations. The response must be as rapid as is reasonable while at the same time maintaining a high level of safety. The knowledge of all members regarding the area and routes of travel is of utmost importance.

3.6 Altered Response Routes

3.6.1 When apparatus is available away from quarters, altered response routes must be considered. The officer shall voice this information to alert incoming units of possible changes in company assignments.

3.6.2 When a response route is changed, be aware of the apparatus responding from normal routes of travel.

3.6.3 Be alert at intersections for other responding units. If units are responding from a location other than home quarters, the officer shall alert other responding companies via radio stating from the location from which they are responding, i.e. "Engine 249 from 51's quarters."

4 FIRE SCENE OPERATIONS

4.1 Objectives

4.1.1 The objectives of any firefighting operation are to protect life and property by performing rescues, protecting exposures, confining the fire, and extinguishing the fire. Water must be applied to the fire in conjunction with the simultaneous venting and primary search operations.

4.2 Strategy

4.2.1 Strategy is the general plan or course of action decided upon to reach the objectives. This is most often indicated by the mode of operation: offensive or defensive.

4.3 Tactics

4.3.1 Tactics are the operations or actions to implement the strategy of the officer in command.

4.4 Size-up

4.4.1 Size-up is an ongoing evaluation of the problems confronted within an emergency situation. Size-up starts with the receipt of the alarm and continues until the fire is under control. This process may be carried out many times and by many different individuals during the incident. The responsibility initially lies with the officer of the first company on the scene. However, all members on an incident must be constantly sizing-up the situation based upon their assignments and responsibilities.

4.4.2 Strategic factors that must be considered during size-up include:
- Time of day
- Life Hazard
- Area (square footage per floor as well as total square footage)
- Height (total building height and height of the fire within the building in relation to the street)
- Construction type
- Occupancy
- Location and extent of fire
- Water supply
- Street conditions
- Auxiliary appliances
- Weather
- Apparatus and equipment (responding and available to respond)
- Exposures (interior and exterior)
- Your resources
- Hazardous materials

4.4.3 If in a multiple-story occupancy, either residential or commercial, check the floor below the fire floor for the layout and location of the reported fire.

4.4.4 The engine company officer must make the decision as to which actions will benefit fireground operations the most. These actions could include rescue, confinement, or extinguishment. Every effort must be made to coordinate the attack with other operations, especially water supply and ventilation. A good rule to follow is:
- R – Rescue
- E – Exposures
- C – Confinement
- E – Extinguishment
- O – Overhaul

4.5 Rescue

4.5.1 Engine companies are often confronted with life saving operations upon arrival. Undoubtedly, it is the most serious factor at any fire. Life-saving operations are placed ahead of firefighting when firefighters are not available to do both (no truck or rescue company on the scene). The best life saving measure may be a prompt attack on the fire which, if allowed to spread, would trap occupants. Life hazard, visible upon arrival, has to be addressed.

4.5.2 Factors entering into rescue decisions are:

- Are occupants in the immediate vicinity of the fire and in danger?
- How many people are trapped?
- Are occupants threatening to jump?
- Are they above the fire?
- Are exits cut off by fire?
- Can they be removed via portable ladders?
- RISK vs. GAIN: Is it a true rescue or is it body recovery?

4.5.3 Actions the engine officer can take to protect the victims are:
- Send firefighters to remove occupants using portable ladders or building exits.
- Protect interior stairways.
- Get hose lines between fire and occupants.
- Extinguish the fire.
- Initiate ventilation to draw fire, heat, and smoke away from occupants.
- Provide assurance to occupants by verbal contact and initiating fire tactics.

4.6 Locating the fire

4.6.1 An exterior survey of the structure and immediate area must be made upon arrival. If the size of the building permits, a lap should be taken to observe as much of the building as possible. Otherwise, the engine officer must solicit reports from units whose vantage point allows a view of the other sides of the structure. If a truck or rescue company is on the scene and the location of the fire is not obvious, the engines should consider using those units to locate the fire and identify the best routes for hose line deployment. The following information must be determined or considered:
- Location of fire or smoke
- The building area and height
- The building entrance(s) and exit(s) such as fire stairs, fire escapes, etc.
- Exposures
- Possibility of obtaining building information from evacuees

4.6.2 An interior survey of the structure must be made for the following:
- Visible fire, smoke, and odors
- Sensed heat or the sound of fire (crackling, etc.)
- Occupants who may provide information
- Stairways leading to that part of the building in which the fire is known or believed to be located
- Structural stability of the interior
- Secondary means of egress

4.6.3 When the initial size-up is completed and the fire located, all pertinent information, and as necessary initial task assignments should be communicated to the affected companies.

4.6.4 There may be cases where lines should not be stretched until the location of the fire has been assured. Premature stretching can take place in the wrong street, into the wrong building, or up the wrong stairs or stairway.

4.6.5 If the engine is on the scene alone, it may be beneficial for the officer and one firefighter to proceed inside to locate the fire, leaving the driver and other firefighter to stretch the line once the fire is found.

4.7 Confinement

4.7.1 Confining the fire means to restrain or prohibit the fire from extending beyond the area involved upon arrival. Confinement of a fire must take into consideration the intensity of the fire as well as the anticipated or known direction of fire travel. In some situations, the mere closing of a door or window may act to confine the fire and permit life saving operations while lines are being stretched. An engine company officer must keep in mind that effective ventilation can help to confine the fire and limit its spread.

4.7.1.1 All members must understand that great care shall be exercised when ventilating so as not to cause unwanted fire extension that might hamper the initial attack line. Good communications are important for a coordinated fire attack. This is particularly true for engines pressing the attack and trucks supporting the attack with ventilation.

4.7.1.2 Exterior fire spread. Officers must be cognizant of the potential fire extension via the exterior to other areas of the structure. This can be one of the fastest means of fire extension. This is the result of fire venting to the outside through a window, door, or other opening, and re-entering the structure through the vented eaves or windows above. In areas where structures are in close proximity to one another, the problem may be even greater in that we now have fire extension to an adjacent building as well. There are several factors contributing to this problem:

- Combustible siding materials
- Distance between dwellings; newer single-family dwellings may be less than 10 feet apart and in some cases as little as 3 feet
- Windows above the venting fire which allow for auto-extension (leapfrogging to floor above)
- Vents in eave lines which lead directly into the attic or cockloft area
- Combustible roofing materials

4.7.1.3 A tactical consideration for addressing exterior fire spread includes a quick sweep of the eaves or siding involved prior to entering the structure. The objective is to drench the siding in order to delay and prevent vertical fire spread. This will delay extension into the exposure. Hose streams should not be directed into windows that are venting fire. Doing so will cause the fire to be pushed into the occupancy.

4.7.1.4 Aggressive firefighting usually means the first engine on the scene will pull the initial attack line to extinguish the fire. However, the first line deployed may not, and in some cases should not, be the line that actually conducts the initial attack. It may be necessary to use the initial line to confine the fire, protect exposures, and protect means of egress. In special circumstances, a number of other tasks may initially take precedence over fire attack.

4.8 Extinguishment

4.8.1 In order to properly affect prompt and efficient extinguishment in as safe a manner as possible, all firefighters must have a thorough knowledge of fire behavior. This includes the "phases of burning" and how they relate to various engine company operations.

4.8.1.1 Particular care must be exercised in dealing with fires that are beyond the transition between free burning and the smoldering phase.

4.8.2 Members must be aware that, due to the tremendous amount of stored heat energy found in many man-made products today, the potential for rapid fire growth as well as "rollover" or "flashover" is great. Therefore, firefighters must be constantly aware of changes in the fire environment during all operations. Engine companies should make every attempt to cool the ceiling and walls with water application utilizing a straight stream pattern to prevent flashover or rollover. Firefighters shall have a thorough knowledge of the various types of attack (direct, indirect, and combination) and when to use each type.

4.8.2.1 Direct Attack – The attack crew advances into the fire area and utilizes a straight or solid stream. The stream is applied directly onto the burning materials until the fire darkens down. This is the attack method most often used for interior firefighting. If necessary, straight or solid streams should be directed at the ceiling to cool the overhead, before directing the stream toward the fire itself in a "Z" pattern. Before advancing, a sweep of the floor with the stream will cool embers and other hot material that could cause firefighter injury.

4.8.2.2 Indirect Attack – The crew attacks the fire from a doorway, window, or other protected area not entering the fire area. A narrow to wide-angle fog stream is directed into the fire room or area. The super-heated atmosphere will turn the water fog into steam. This method of extinguishment absorbs the heat and displaces the oxygen. <u>This method of extinguishment **must not** be used in areas where victims may be located or firefighters are operating</u>.

4.8.2.3 Combination Attack – This method utilizes both the direct and indirect methods at the same time. A narrow fog stream is used in a "T," "Z," or "O" pattern. The fog in the higher atmosphere turns to steam and the fog at floor level hits the burning material. <u>As with an indirect attack, this method **must not** be used in areas where victims may be located or firefighters are operating.</u>

4.8.3 Initial actions of hose lines should be to sweep the ceiling and walls to prevent flashover or rollover if needed. If the hose line cannot advance due to <u>high heat conditions</u>, the officer should call for *additional ventilation*. If the advance is being hindered by <u>heavy fire</u>, *additional lines or larger lines should be ordered*.

4.9 Determining Fire Flow

4.9.1 There are several means for determining fire flow, all of which are based upon the amount of heat released from materials as they burn and the amount of heat that water absorbs while raising its temperature to that of vaporized steam. It must be recognized that these are general guidelines for determining the *amount* of water to be applied. Experience and logic must be employed in determining how the water should be applied.

4.9.2 Care should be taken not to characterize fire loads based solely upon occupancy. It should be remembered that not all occupancies are used for their designed purpose. Outward appearances can be deceiving. For example, excessive storage of combustible liquids or other materials in a single-family structure would obviously change the fire loading within the structure.

4.9.3 <u>Light fire loads require a flow of 10 GPM per 100 square feet of involved area.</u> This flow rate can most often be delivered using 1 ¾-inch hose lines.

4.9.3.1 A typically furnished home or apartment may be classed as a "light" fire load.

4.9.4 <u>Medium fire loads require a flow of 20 GPM per 100 square feet of involved area.</u> Use of 2½-inch hose lines should be expected.

4.9.4.1 "Medium" fire loads would be characterized as commercial occupancies containing moderate amounts of combustible materials.

4.9.5 <u>High fire loads require a flow of 30 GPM per 100 square feet of involved area.</u> Use of 2 ½-inch hose lines or heavy-caliber master streams will be needed.

4.9.5.1 "High" fire loads would be characterized as storage facilities that house large quantities of combustible material or contents that are capable of unusually high rates of heat release.

4.10 Rules for Stream Positioning

4.10.1 It must be understood that more lives can be saved from fires by proper stream positioning than by any other fireground operation.

4.10.2 If the fire cannot be quickly confined and extinguished, then the following general rules for positioning streams shall apply:

- When human life is endangered, the first stream is placed between the fire and persons endangered by it
- When human life is not endangered, then the first stream is placed between the fire and the most severe exposure

4.10.3 When a back-up line is deployed, it should be stretched to the same point as the first line or as otherwise directed. Normally, the back-up line is capable of equal or greater flow than the original line. However, a back-up line, at a minimum, should be capable of delivering the same amount of GPM as the original line.

- This hose line backs up the first in the event the first stream is inadequate or malfunctions. The second line is for support of the first line and to help ensure success
- This line must have enough hose to cover the floor above the fire floor
- Firefighters must not play two streams against each other
- If possible, attack should be made from the unburned side; however, this is often not practical.
- No more than two hose lines should be deployed through any one doorway.

1.1.4 When more than one line is deployed from a single engine, it is recommended to have a water supply established before the second line is charged. However, in almost every case, the attack engine is supplied by the 2nd due engine and the crew from the second engine deploys the second line. Water supply should be established concurrently with the deployment of the second line. The exception to this would include the "deluge blitz attack" where the 1st arriving engine opts to operate their master stream to knock down the bulk of the fire before a water supply is established.

1.1.5 Once water supply is established to any attack engine, the number of lines stretched from that engine is only limited by the amount of incoming water.

1.1.6 Additional lines may be stretched depending upon occupancy or fire situation. They may be deployed to:

- Cover secondary means of egress
- Protect persons trapped above the fire
- Proceed to adjoining buildings to protect exposures or operate through common horizontal openings
- Prevent and extinguish the vertical spread of fire
- Attack from the unburned portion of the structure and protect the stairway
- Other priorities as determined by the incident commander

1.1.7 It must be understood that in order to coordinate ventilation operations with fire attack, the engine must convey to the outside vent team that they are ready to advance. Ventilation openings should be made prior to fire stream application. This communication can be made either by radio or face–to-face.

4.10.4 Members should consider the following points for proper stream application.

4.10.4.7   If it becomes necessary to conduct fire attack from the exterior of the structure, you must take into account members who might be operating inside the structure, i.e. search or ventilation crews. Additionally, any victims still in the structure must be considered. Again, effective communication via radio is of paramount importance in these situations.

4.10.4.8   When a direct attack cannot be made from the interior due to adverse wind conditions or excessive heat, consideration should be given to breaching the wall by making a hole in an adjoining room or compartment and conducting the initial attack in that way. In all cases, great care must be exercised in protecting the hallway or other means of egress. Maintaining the door to the fire room in the closed position until ready for interior attack will most likely be your best means of accomplishing this objective.

4.10.4.9   Fires that are free burning require a straight or solid stream to be effective. This technique is called the direct method of attack. A direct attack will be utilized on most interior fires. In order for the direct attack to be successful, coordination with the other units on the scene who are venting and searching must be considered. The direct method would <u>not</u> be utilized when fire is in areas that are unoccupied and areas which due to intense heat conditions cannot be entered. Examples are attics, knee-walls, and other confined areas.

4.10.5   To determine the type of attack that should be utilized, the following factors need to be taken into consideration:

- Fire attack crews have access to the seat of the fire
- If other firefighters need to be in the fire area, such as performing primary searches
- Victims and firefighters present or suspected to be in the fire area that could be affected by steam production
- Inadequate ventilation that would not allow the smoke and hot gases to escape to the outside
- If all of the above items exist, the direct method of attack should be used

4.11   Engine Officer

4.11.1   Once the decision to advance a hose line has been determined, a properly trained engine company should be able to do the following without close personal supervision of the officer:

- Position the apparatus as necessary to effect an adequate water supply
- Estimate the amount of hose needed for the stretch
- Remove the hose from the apparatus
- Stretch the proper length of hose line to reach the objective
- Connect and utilize any appliances that may be needed

4.11.2   It is the first-due engine's responsibility to lay the supply line(s). When possible, the first-due engine company should forward lay the supply line. However, in some situations this may be impractical and a reverse or split lay may be utilized. The officer must make a proper decision to lay single or multiple lines. The old adage, "WHEN IN DOUBT, LAY IT OUT" seems to apply. It is much easier to lay dual lines initially than to try to hand lay a second one later.

4.11.3   If there is a known rescue situation, the officer should consider whether or not to stop at the water source and lay out, or proceed directly to the scene to avoid any delay to attempt rescue.

4.11.4   While the attack line is being stretched, the engine officer must be gathering information relative to the location and extent of the fire. The gathering of this information normally takes relatively little time. The advantages of proper line placement and rapid line advance are well worth the effort. The officer's size-up and tactical considerations for decision-making should include the following items:

4.11.4.1 Monitoring *all* radio transmissions.

4.11.4.2 The crew may need to locate the fire and search for victims without the aid of a hose line if no ladder or rescue company is present to accomplish this task.

4.11.4.3 In a non-standpipe equipped multiple story occupancy, the engine officer must use good judgement in determining whether to stretch the attack line wet or dry.

4.11.4.4 The engine officer should communicate to the nozzle team any information that might be critical, i.e. "straight down the hall, second door on the left." He should then ensure that the nozzle team is ready and initiate the advance. It is the engine officer's responsibility to ensure persons, either firefighters or civilians, are not adversely affected by the application of the fire stream. If there is *any doubt* as to whether there are victims or firefighters on the opposite side of the fire from where the attack is made, then a direct attack, utilizing a straight or solid stream, shall be used.

4.11.4.5 While the attack line is being advanced, the engine officer should monitor all radio transmissions. This affords him knowledge of any hazardous condition that may be developing, i.e. ventilation problems, fire extension, collapse potential, and "Mayday" transmissions. It is understood that this may not be possible under the noise conditions of the typical fire attack. Therefore, the engine officer should, when possible, initiate communication and expect acknowledgment. The engine officer, in initiating communications, will be providing important information for the incident commander and other officers and units on the scene. These messages should be short and concise, i.e. fire knocked down. They shall include current conditions inside the structure as well as any resources that might be needed.

4.11.4.6 The engine officer should constantly monitor progress and conditions. He directs the nozzle team's advance and instructs them as to their next objective.

4.11.4.7 Firefighters should avoid the fire floor and unprotected areas where active fire is present without an attack line.

4.11.4.8 The engine officer should constantly be monitoring the welfare of the nozzle team, being vigilant of anything or condition that might unnecessarily threaten their safety. He must be aware of the tendency of most firefighters to want to stay inside, well past the point when they should be relieved, and take action necessary to get other personnel to relieve them.

4.11.4.9 The engine officer should relay orders in concise and clear tones not higher than normal, dependant upon conditions, and in as few words as possible. He should keep his nozzle team informed of progress, giving estimates of what remains to be extinguished. It must be realized, due to the inevitable noise generated at a working incident, normal talking will be difficult to hear.

4.11.4.10 The engine officer *should* be with the nozzle or as close as possible. Due to the cramped quarters found at many incidents, the engine officer may find himself in the position of back-up person.

4.11.4.11 As the crew advances, the engine officer should be checking for fire extension in areas immediately adjacent to their position, such as adjoining rooms. Additionally, the officer must be looking for any victims that may be in the immediate area and always monitoring the structural conditions and fire behavior in their area of operation.

4.12 Booster lines are not normally considered for interior operations, other than to standby during overhaul operations to assist the investigator in cooling down an occasional hot spot.

4.13 Hose Line Advancement

4.13.4.1 The following should be considered when choosing the correct length of hose for deployment:
- Engine position and distance to the building, known as "setback"
- Height of the fire floor
- The area of the building and the location of the fire within that area
- Amount of hose needed to reach the floor above the fire if the need arises
- Attack line should be long enough to have at least one length available at the point the attack commences. (One length meaning 50 feet of hose).
- One method of calculating the proper length of hose in commercial buildings: take the setback + length of the building + width of the building + one length for each floor above the engine + one length at the point of attack = proper length of hose needed. Note that this formula would not apply to single-family dwellings.

4.13.4.2 Guidelines To Assist with Smooth Line Advancement

4.13.4.3 Chock doors open!

4.13.4.4 Ensure uncharged lines advanced into a structure do not become wedged under a partially closed door.

4.13.4.5 Stretching a line up a stairwell using the well-hole is very effective and uses only about one length per floor. However, if a crew is stretching up a narrow well-hole, ensure the line does not become wedged between the railing and the steps. Have additional members in the stairwell to advance the line into the hallway.

4.13.4.6 Limit the number of lines to no more than two through any one entrance. If additional lines are required, they should be advanced through alternate doors, windows, upper floors, over ladders or by hoisting with ropes.

4.13.4.7 Use of the aerial as a permanent alternate standpipe is not generally advised. It should be free for other tasks, such as access to upper floors and the roof. More importantly, the aerial should be available to move into a position for emergency egress for firefighters or victims that show at a window or balcony. However, the aerial can be used to move hand lines into position.

4.13.4.8 When working out of a stairwell, stretch the line to the upper landing to ease advancement onto the fire floor from the stairwell. The uncharged hose should be laid out in accordion style to ease advancement.

4.13.4.9 At times, the best point of entry for the attack line might have to be accessed by advancing the line up and over a ladder or by entering the building and then hoisting the line.

4.13.4.10 Secure hose lines to the railings or windows on upper floor landings to prevent injuries to firefighters and occupants. This keeps stairwells free of clutter and prevents the hose from kinking.

4.13.4.11 When preparing to advance a hand line down a basement stairwell, ensure there is enough line poised at the top of the stairway to make a smooth unobstructed advance into the lower floor level.

4.13.4.12 When stretching lines inside a structure, ensure that firefighters are positioned at intervals over the length of the line in order to keep the line moving.

4.13.4.13 When multiple lines are required, consider utilizing more than one engine for attack positions. The entire operation would not be dependent upon a single attack engine if that engine were to malfunction. Reliance upon a single source of water supply should be avoided.

4.13.4.14 When the 2½-inch hose line is deployed it will take two crews to properly advance this hose line into a structure. Firefighter positions on the line should have the nozzle person and back-up at the nozzle and a member at approximately each coupling.

4.13.4.15 Rural Water Supply

4.13.4.16 In non-hydrant areas, engine companies have to rely on alternative water sources or utilize remote hydrants. Water supply is achieved by utilizing either a relay operation or a water shuttling operation.

4.13.4.17 The standard relay operation will place an engine approximately every 800 feet from the engine providing the fire attack.

4.13.4.18 A standard relay would:

4.13.4.19 Begin with an attack engine at the fire scene. (In restricted narrow access areas the attack engine will lay dual 3-inch supply lines or Large Diameter Hose (LDH) from the street and proceed to the fire).

4.13.4.20 A second engine would reverse or split-lay dual 3-inch supply lines (800 feet) or LDH toward the water source.

4.13.4.21 This sequence would continue until the water source is met and established ensuring an engine approximately every 800 feet.

4.13.4.22 Tanker shuttle

4.13.4.23 A guideline to use in deciding upon a relay or a shuttle is that if a relay requires more than 3 engines, in addition to the attack engine, a tanker shuttle should be considered.

4.13.4.24 Begin with an attack engine at the fire scene. (In restricted, narrow access areas, the attack engine will lay supply lines from the street and proceed to the fire).

4.13.4.25 The tanker will supply the attack engine as the portable folding tanks are set up at an accessible location, reasonably close to the attack engine, such as at the end of the driveway.

4.13.4.26 The supply engine will draft from the static source at the folding tanks and supply the attack engine. The supply engine shall position to allow complete access and egress for the tanker operations.

4.13.4.27 The tanker will begin to fill the portable tanks. In order to ensure that the supply engine has the capability to draft, one folding tank will only be half filled. When the supply engine obtains a draft, the tanker will completely fill the portable folding tanks.

4.13.4.28 The tanker would remain mobile and proceed to the closest reliable water source, refill, and shuttle water back to the folding tanks.

4.13.4.29 Additional engines and tankers could be added to the shuttle operation. If additional engines are used to shuttle water, a 100-foot supply line or lines should be placed near the folding tanks to be utilized by the shuttle engines to off-load. This will allow access for the tanker to operate without obstacles.

4.13.4.30 In order to fill the shuttle apparatus, an engine will be required to establish a water supply and fill site.

4.13.4.31 If the tanker is delayed or responds from a remote location, alternate supply methods need to be deployed.

4.13.4.32 When either a relay or shuttle operation is required, a water supply officer must be established.

4.13.4.33 To avoid obstructing the access of the shuttle operation where access to the structure is limited, it is advisable for the truck company to remain on the street and hand-carry equipment to the scene.

Appendix C
NIOSH: Death in the Line of Duty Report

Death in the line of duty...

A summary of a NIOSH fire fighter fatality investigation   May 16, 2008

Career Fire Fighter Dies in Wind Driven Residential Structure Fire – Virginia
Revised June 10, 2008 to clarify Recommendation #2

SUMMARY

On April 16, 2007, a 24-year-old male career fire fighter (the victim) was fatally injured while trapped in the master bedroom during a wind-driven residential structure fire. At 0603 hours, dispatch reported a single family house fire. At 0609 hours, the victim's ladder truck was second to arrive on scene. Fire was visible at the back exterior corner of the residence. Noticing cars in the driveway, no one outside, and no lights visible in the house, the lieutenant from the first arriving engine called in a second alarm. A charged 2 ½" hoseline was stretched to the front door by the first arriving engine crew. The engine crew stayed at the door with the attack line while the cause of poor water pressure in the hoseline was determined. The victim and his lieutenant, wearing their SCBA, entered the residence through the unlocked front door. With light smoke showing, they walked up the stairs to check the bedrooms. The victim and lieutenant cleared the top of the stairs and went straight into the master bedroom. With smoke beginning to show at ceiling level, the victim did a right-hand search while the lieutenant with thermal imaging camera (TIC) in-hand checked the bed. Suddenly the room turned black then orange with flames. The lieutenant yelled to the victim to get out. While verbal communication among the crew was maintained, the lieutenant found the doorway and moved toward the stairs. He ended up falling down the stairs to a curve located midway in the staircase. The lieutenant tried to direct the victim to the stairs verbally and with a flashlight. As the wind gusted up to 48 miles per hour, the wind-driven fire and smoke engulfed the residence. The incident commander (IC) ordered an evacuation and the lieutenant was brought outside by the engine and rescue company crews. The ladder truck lieutenant received burns on his ears and right index finger. At 0614 hours, the rescue company officer issued a Mayday followed by the victim's Mayday. With protection from hose lines, several attempts were made by the engine and rescue company crews to reach the second floor. On the third attempt the stair landing was reached but the ceiling started collapsing and flames intensified. At 0621 hours, due to the intensity of the fire throughout the structure, all fire fighters were evacuated, operations turned defensive, but the incident continued in rescue mode. At 0657 hours, the victim was found in the master bedroom partially on a couch underneath the front windows.

The Fire Fighter Fatality Investigation and Prevention Program is conducted by the National Institute for Occupational Safety and Health (NIOSH). The purpose of the program is to determine factors that cause or contribute to fire fighter deaths suffered in the line of duty. Identification of causal and contributing factors enable researchers and safety specialists to develop strategies for preventing future similar incidents. The program does not seek to determine fault or place blame on fire departments or individual fire fighters. To request additional copies of this report (specify the case number shown in the shield above), other fatality investigation reports, or further information, visit the Program Website at www.cdc.gov/niosh/fire/ or call toll free **1-800-CDC-INFO** (1-800-232-4636).

Fatality Assessment and Control Evaluation
Investigation Report # F2007-12

Career Fire Fighter Dies in Wind Driven Residential Structure Fire – Virginia

NIOSH investigators concluded that, to minimize the risk of similar occurrences, fire departments should:

- *ensure that standard operating procedures (SOPs) for size-up and advancing a hoseline address the hazards of high winds and gusts*
- *ensure that primary search and rescue crews either advance with a hoseline or follow an engine crew with a hoseline*
- *ensure that staffing levels are sufficient to accomplish critical tasks*
- *ensure that fire fighters are sufficiently trained in survival skills*
- *ensure that Mayday protocols are reviewed, modified and followed*
- *ensure that water supply is established and hoses laid out prior to crews entering the fire structure*
- *ensure that fire fighters are trained for extreme conditions such as high winds and rapid fire progression associated with lightweight construction*

Additionally, municipalities should:

- *ensure that dispatch collects and communicates information on occupancy and extreme environmental conditions*

Although there is no evidence that the following recommendation could have specifically prevented this fatality, NIOSH investigators recommend that fire departments:

- *ensure that radios are operable in the fireground environment*

INTRODUCTION

On April 16, 2007, a 24-year-old male career fire fighter (the victim) was fatally injured while trapped in the master bedroom during a residential structure fire. On April 17, 2007, the fire department, U.S. Fire Administration (USFA), and the International Association of Fire Fighters (IAFF) notified the National Institute for Occupational Safety and Health (NIOSH) of this fatality. On May 6 - 9, 2007, a General Engineer and a Safety and Occupational Health Specialist from the NIOSH Fire Fighter Fatality Investigation and Prevention Program investigated the incident. Photographs of the incident scene were taken and meetings were conducted with the Battalion Chief of Health and Safety (fire department's Investigating Team Leader), Fire Marshal, and an IAFF representative. Interviews were conducted with officers and fire fighters who were at the incident scene. The NIOSH investigators reviewed the department's standard operating guidelines (SOGs), the officers' and victim's training records, photographs of the incident scene, written witness statements, the coroner's report, and a weather station report. At the request of the fire department, NIOSH examined and evaluated the

Fatality Assessment and Control Evaluation
Investigation Report # F2007-12

Career Fire Fighter Dies in Wind Driven Residential Structure Fire – Virginia

victim's SCBA. The SCBA was examined component by component to determine conformance to the NIOSH approved configuration. The SCBA was too damaged to test performance. (see Appendix)

Fire Department

The combination department has nineteen fire stations, three administrative worksites, a warehouse, and a training facility. A total of 1,478 fire and rescue personnel (452 career and 1,026 volunteer) serve a population of about 384,000 residents in a geographic area of about 348 square miles.

Personal Protective Equipment

At the time of the incident, the victim was wearing personal protective equipment consisting of turnout coat and pants, gloves, a helmet, hood, SCBA with an integrated PASS device, and he carried a radio. Given the condition of the victim's SCBA, the NIOSH post-incident evaluation could not determine if the SCBA performance contributed to the fatal incident. (see appendix).

Apparatus and Personnel

Dispatch reported a single family house fire at 0603 hours.

On scene at 0608 hours:

Engine #12 [E12] – Lieutenant (LT#1), engine operator, and a fire fighter

On scene at 0609 hours:

Truck #12 [T12] – Lieutenant (LT#2), truck operator, and two fire fighters (one the victim)

FireMedic #12 [M12] – Lieutenant and a fire fighter

On scene at 0610 hours:

Rescue #10 [R10] – Lieutenant, driver operator, and three fire fighters

Engine #10 [E10] – Technician II (Acting Officer), engine operator, and a fire fighter

On scene at 0611 hours:

Battalion Chief (Incident Commander (IC))

On scene at 0612 hours:

Engine #20 [E20] – Captain, engine operator, engine operator in training, and two fire fighters

Ambulance #10 [A10] – Lieutenant and two fire fighters

Page 3

**Fatality Assessment and Control Evaluation
Investigation Report # F2007-12**

Career Fire Fighter Dies in Wind Driven Residential Structure Fire – Virginia

Safety #02 - Safety Officer

On scene at 0618 hours:

Engine #2 [E2] – Lieutenant, engine operator, and two fire fighters

Training/Experience
The victim had completed National Fire Protection Association (NFPA) Fire Fighter Level I and II training, Cardiopulmonary Resuscitation (CPR), Critical Incident Stress Management, Hazmat Operations, Fire Fighter Survival Skills I & II, Infection Control and several other technical courses. The victim was a career fire fighter with one year of fire fighting experience.

The Incident Commander had completed Intermediate and Advanced National Incident Management System training, Intermediate and Advanced Incident Command System Courses, several HAZMAT courses, and various other administrative, personnel and technical courses. The Incident Commander is a career fire fighter with 24.5 years in the fire service at the time of the incident.

Lieutenant #1 (LT#1) had completed Fire Fighter 1 and 2, Fire Officer 1, Incident Officer, several HAZMAT courses, Fireground Tactics, and various other administrative and technical courses. LT#1 is a career fire fighter with 8 plus years in the fire service at the time of the incident.

Lieutenant #2 (LT#2) had completed Fire Fighter 1 and 2, Fire Officer 1 and 2, Fire Fighter Survival Skills 1 and 2, Advanced Fire Fighter Safety Skills, several HAZMAT courses, and various other administrative and technical courses. LT#2 is a career fire fighter with 8 years in the fire service at the time of the incident.

The Safety Officer had completed Fire Fighter 1, 2, and 3, Fire Officer 1 and 2, Field Officer, Incident Officer, Fire Fighter Survival Skills 1 and 2, several HAZMAT courses, and various other administrative and technical courses. The Safety Officer is a career fire fighter with a total of 16.5 years in the fire service at the time of the incident.

Building Information
The building was an approximately 6000 square foot, two-story plus finished walkout basement, non-sprinklered residential structure that was constructed of wood framing with vinyl siding and a brick veneer front on the exterior. The residence had a large 700 square foot wood deck that ran three quarters the length of the rear of the structure on the first floor (C-side). The roof consisted of wood rafters with fiberglass shingles over oriented strand board sheathing (see Photo1).

NIOSH Fire Fighter Fatality Investigation and Prevention Program

FATALITY Assessment and Control Evaluation Investigation Report # F2007-12

Career Fire Fighter Dies in Wind Driven Residential Structure Fire – Virginia

Photo 1: A-side of the structure where the victim entered. The victim was found in the second floor master bedroom above the bay window.

Weather
At the time of the incident, the conditions were overcast with an approximate temperature of 45 degrees Fahrenheit and a measured sustained wind speed of 25 miles per hour (mph) from the Northwest with wind gusts up to 48 mph.

INVESTIGATION
On April 16, 2007, a 24-year-old male career fire fighter (the victim) was fatally injured while trapped in the master bedroom during a wind-driven residential structure fire. At 0603 hours, dispatch reported a single family house fire. At 0608 hours, the victim's ladder truck (T12) was second on the scene directly behind engine 12 (E12). The crews encountered fire in the B/C exterior corner of the residence underneath and along the deck on the first floor on the C-side on the structure. The wind was blowing at speeds of 25 to 48 mph. The residence was located at the top right side of a drainage

Page 5

**Fatality Assessment and Control Evaluation
Investigation Report # F2007-12**

Career Fire Fighter Dies in Wind Driven Residential Structure Fire – Virginia

where the wind was directed onto the C-side of the house. The lieutenants from E12 and T12 (LT#1 and LT#2, respectively) walked around opposite sides of the structure and met in front to discuss the size-up. Noticing cars in the driveway, no-one outside, and no lights visible in the house, LT#1 called in a second alarm. *(Note: A neighbor drove to the residence and woke the residents. He walked them to another neighbor's house and walked back to move his vehicle in anticipation of the fire department's arrival. No-one relayed to the fire department that the residents were out of the house until after interior operations were underway.)*

A charged 2 ½" hoseline was stretched to the front door from E12 by the E12 crew. LT#2 and the victim donned their SCBA while waiting to enter the structure. The victim tried the door, which was unlocked, so the victim and LT#2 walked into the foyer. E10 arrived on scene and was instructed by LT#1 to pull 300 feet of 1 ¾" hose from E12 in case it would be needed in the rear. At 0611 hours, with a light haze of smoke visible on the first floor, the T12 crew walked up the stairs to check the bedrooms. The E12 crew was delayed at the door with the attack line due to poor water pressure. LT#1 straightened out several kinks in the hose prior to entering the structure. At the top of the stairs, LT#2 and the victim encountered smoke banked down 3 feet from the second floor ceiling and went to the D-side of the residence towards the master bedroom. At 0611 hours, a Battalion Chief arrived on scene and assumed incident command. E20 arrived on scene and was instructed by the IC to pull 200 feet of 1 ¾" hose from E12 and cover the D-side exposures. (see Diagram#1)

LT#2 and the victim came to a set of double doors. The right door was open and they entered the master bedroom. With smoke showing at ceiling level, the victim did a right-hand search and LT#2 with thermal imaging camera (TIC) in-hand checked the bed. Suddenly the room turned black then orange with flames. LT#2 yelled to the victim to get out. While verbal communication among the crew was maintained, LT#2 found the doorway and crawled towards the stairs, falling to the curve located midway in the staircase. LT#2 communicated verbally and visually (via a flashlight) in an attempt to direct the victim to the stairs. LT#2 had been burned on his ears and right index finger. *(Note: Personal protective equipment (PPE), such as hood and gloves were properly worn. However, the helmet liner was not properly down over the ears. The PPE received direct flame impingement and heat exposure.)*

LT#1 and his crew along with the R10 crew were still at the front door which had slammed closed. When the door was re-opened, fire engulfed the doorway and LT#1 started yelling for LT#2 and the victim to come down the stairs. LT#1 noticed the severe change in the fire conditions in the stairway and requested an emergency evacuation. At 0613 hours, the wind-driven fire and smoke engulfed the residence. Fire was coming out of the eaves on the D-side of the structure. The E20 crew was flowing water on the D-side and the E10 crew was flowing water on the A-side of the residence, but the wind hampered the attack. Several ladders were thrown on the A and D-sides to the second floor, and on the C-side deck to the first floor, but the high winds and extreme heat made it difficult to stabilize the ladders against the building.

Page 6

NIOSH Fatality Assessment and Control Evaluation Investigation Report # F2007-12

Career Fire Fighter Dies in Wind Driven Residential Structure Fire – Virginia

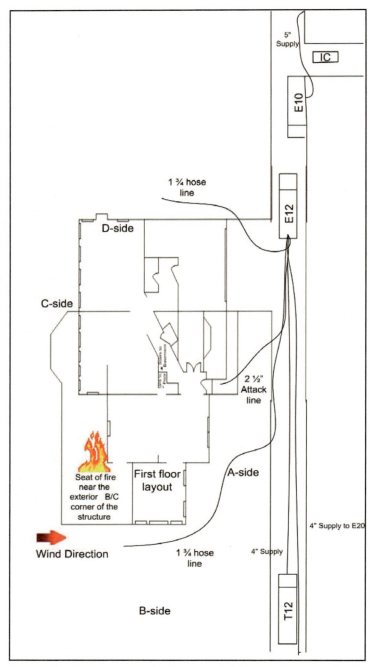

Diagram #1: Apparatus and hoseline location at time of incident

**Fatality Assessment and Control Evaluation
Investigation Report # F2007-12**

Career Fire Fighter Dies in Wind Driven Residential Structure Fire – Virginia

The engine operator from E12 blew the air horn for 10 seconds to signal an evacuation. The IC gave a follow-up order to evacuate over the fireground channel. Concurrently, the R10 crew located LT#2 in the staircase area and brought him outside. At 0614 hours, LT#2 informed them that the victim was near the stairs. The R10 lieutenant issued a Mayday. It was shortly followed on the radio by the victim's Mayday. Two lines, a 2 ½" and a 1 ¾", were flowing on the A-side of the structure with minimal impact. *(Note: From the beginning of the incident, low water pressure in two hoselines was an issue. Removing kinks in the hoses helped somewhat, but pressure problems persisted. It was undeterminable if the resultant low pressure was at the hydrant, engines, and/or due to hoseline deployment.)*

At 0615 hours, LT#1 and a R10 fire fighter made it to the top of the stairs, but heat pushed them down. Several attempts were made by the E12 and R10 crews to go back up to the second floor. On the third attempt to ascend the staircase, and the second time the landing was reached, the ceiling started collapsing and flames intensified.

At 0621 hours, the Safety Officer, seeing the intensity of the fire throughout the structure, instructed the IC to call for another evacuation. The engine operator from E12 blew the air horn for a second time. At this point, the incident turned to a defensive attack, but rescue mode continued. At 0621 hours, the E2 crew was designated as the RIT and entered the C-side of the structure with a 1 ¾" hoseline to search for the victim in case a floor collapse had occurred.

At 0631 hours, additional crews attempted to re-enter and search for the victim, but they were unable to reach the second floor due to intense heat and fire conditions. At 0634 hours, command requested dispatch of a third alarm. At 0643 hours, crews with TICs in hand were able to reach the second floor, but due to structural collapse and high-heat conditions, access to second floor areas was limited. *(Note: The victim's PASS was never heard by fire fighters on the fire ground. Due to extreme heat damage, post incident testing was not possible.)* At 0657 hours, the victim was found in the master bedroom partially on a couch underneath the bedroom A-side windows. (see Diagram #2)

Page 8

Fatality Assessment and Control Evaluation
Investigation Report # F2007-12

Career Fire Fighter Dies in Wind Driven Residential Structure Fire – Virginia

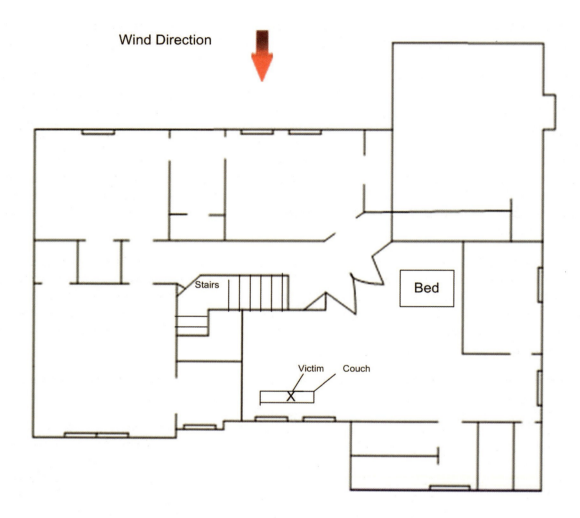

Diagram #2. Location of victim on second floor in master bedroom of fire structure

CAUSE OF DEATH
The coroner listed the cause of death as thermal and inhalational injuries.

Fatality Assessment and Control Evaluation
Investigation Report # F2007-12

Career Fire Fighter Dies in Wind Driven Residential Structure Fire – Virginia

RECOMMENDATIONS

Recommendation #1: Fire departments should ensure that standard operating procedures (SOPs) for size-up and advancing a hoseline address the hazards of high winds and gusts

Discussion: Fire departments should develop SOP's for incidents with high-wind conditions including defensive attack if necessary. Weather can be considered as critically important when at the extreme, and relatively unimportant during normal conditions.[1] Wind has a strong effect on fire behavior which includes supplying oxygen, reducing fuel moisture, and exerting physical pressure to move the fire and heat. Wildland fire fighters are very familiar with these effects of wind on the rate at which fire spreads. According to Dunn, "When the exterior wind velocity is in excess of 30 miles per hour, the chances of conflagration are great; however, against such forceful winds, the chances of successful advance of an initial hose line attack on a structure fire are diminished. The firefighters won't be able to make forward hoseline progress because the flame and heat, under the wind's additional force, will blow into the path of advancement."[2]

Fire fighters should change their strategy when encountering high wind conditions. An SOP should be developed to include obtaining the wind speed and direction, and guidelines established for possible scenarios associated with the wind speed and the possible fuel available, similar to that in wildland fire fighting.[2] When the interior attack line has little or no effect on the fire, the line should be withdrawn and a second hoseline should be advanced on the upwind side of the fire. This method may require the use of an aerial ladder or portable ladder, if safety permits.[2]

Recommendation #2: Fire departments should ensure that the primary search and rescue crews either advance with a hoseline or follow an engine crew with a hoseline

Discussion: Hoselines can be the last line of defense, and the last chance for a lost firefighter to find egress from a burning building. According to the USFA Special Report: *Rapid Intervention Teams and How to Avoid Needing Them*, the basic techniques taught during entry level fire fighting programs describe how to escape a zero-visibility environment using only a hoseline.[3] However, as years elapse from the time of basic training, fire fighters may overlook this technique. Exiting a structure in zero visibility should be simple, fast and easy for a fire fighter with a hoseline. A fire fighter operating on a hoseline should search along the hose until a coupling is found. Once found, the fire fighter can "read" the coupling and determine the male and female ends. The IFSTA manual *Essentials of Fire Fighting* teaches that the female coupling is on the nozzle side of the set and the male is on the water side of the set. In most cases, the male coupling has lugs on its shank while the female does not. Once oriented on the hose, a fire fighter can follow the hoseline in the direction away from the male coupling which will take you toward the exit.[4] There are a number of ways that fire hose can be marked to indicate the direction to the exit, including the use of raised arrows and chevrons that provide both visual and tactile indicators. Fire departments may use a variety of techniques to train fire fighters on how to identify hoseline coupling and the direction to the exit, based on the model of hose used by the department. The key point is that this training needs to be conducted and repeated often so that fire fighters are proficient in identifying the direction to the exit in zero visibility

Page 10

Fatality Assessment and Control Evaluation Investigation Report # F2007-12

Career Fire Fighter Dies in Wind Driven Residential Structure Fire – Virginia

conditions while wearing gloves, the hose entangled, and with various obstructions present. This procedure should be incorporated into SOPs, trained upon, and enforced on the fireground.

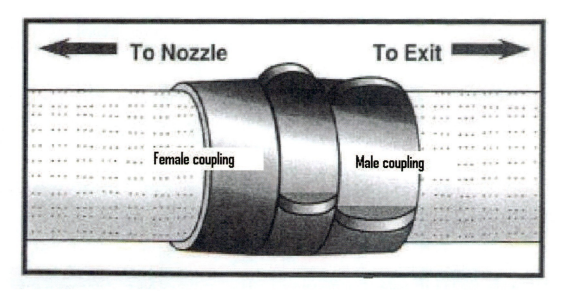

Diagram #3. Hose couplings will indicate the direction toward the exit. Adapted from IFSTA Essentials of Fire Fighting, 4th Edition.

In this incident, the truck crew went into the master bedroom doing a search without a hoseline. The engine crew with the hoseline was still at the front door when the truck crew went into the bedroom. Seconds later heavy smoke and flames blew through the upstairs hallway. The fire department's SOGs allow for the primary search and rescue crew not to have a hoseline as long as they are within sight of a crew with a hoseline. However, situations arise where conditions change in seconds preventing a fire fighter from following the hoseline to safety unless it is immediately available.

Recommendation #3: Fire departments should ensure that staffing levels are sufficient to accomplish critical tasks

Discussion: The National Fire Protection Association (NFPA) 1710 Standard identifies the minimum resources for an effective fighting force to perform critical tasks. These tasks include establishing water supply, deploying an initial attack line, ventilating, performing search and rescue, and establishing a RIT, etc. NFPA 1710 recommends that the minimum staffing levels for an engine company to perform effective and efficient fire suppression tasks is four.[5]

In this case, the first arriving engine (E12) was staffed with a lieutenant, an engine operator and a fire fighter. They stretched a charged 2 ½" attack line to the front door of the involved building for the

Fatality Assessment and Control Evaluation Investigation Report # F2007-12

Career Fire Fighter Dies in Wind Driven Residential Structure Fire – Virginia

initial attack. It is extremely difficult for a two man crew (Lieutenant and fire fighter) to advance or operate a 2 ½" hoseline without assistance. For large diameter hoselines a 3 or 4 man operation significantly increases mobility and efficiency. Had the attack line been a 1 ¾" hoseline, they would have been much better able to rapidly advance a charged line up the stairs.

Recommendation #4: Fire departments should ensure that fire fighters are sufficiently trained in survival skills

Discussion: Fire fighters trapped or disoriented inside a room should be trained to rapidly locate doors and windows in order to escape. This is a skill that every interior structural fire fighter should possess and is typically taught in Firefighter Survival Skills I & II classes. Understanding when to self-rescue, and when to stay in a location to be rescued are critical. Fire departments should provide periodic refresher training to ensure fire fighters can effectively apply this training in different scenarios.

Recommendation #5: Fire departments should ensure that Mayday protocols are reviewed, modified and followed

Discussion: Fire fighters must act promptly when they become lost, disoriented, injured, low on air, or trapped.[6-10] First, they must transmit a distress signal while they still have the capability and sufficient air. The next step is to manually activate their PASS device. To conserve air while waiting to be rescued, fire fighters should try to stay calm and avoid unnecessary physical activity. If not in immediate danger, they should remain in one place to help rescuers locate them. They should survey their surroundings to get their bearings and determine potential escape routes, and stay in radio contact with Incident Command and rescuers. Additionally, fire fighters can attract attention by maximizing the sound of their PASS device (e.g., by pointing it up in an open direction), pointing their flashlight toward the ceiling or moving it around, and using a tool to make tapping noises.

A crew member or other fire fighter who recognizes a fellow fire fighter is missing or in trouble should quickly try to communicate with the fire fighter via radio and, if unsuccessful, initiate a mayday for that fire fighter providing relevant information as described above.

Department protocol requires that when a Mayday is transmitted, the IC must either personally handle the situation or designate another officer to do so. Part of "handling" a mayday is to communicate with the trapped or lost fire fighters and with any other fire fighters or officers involved. The IC or designated officer must communicate the emergency to all fireground personnel to minimize extraneous radio communication and designate another radio channel for normal fireground operations.

Recommendation #6: Fire departments should ensure that the water supply is established and hoses laid out prior to crews entering the fire structure

Discussion: Successful fire suppression and fire fighter safety depends upon discharging a sufficient quantity of water to remove the heat being generated and provide safety for the interior attack crews.

**Fatality Assessment and Control Evaluation
Investigation Report # F2007-12**

Career Fire Fighter Dies in Wind Driven Residential Structure Fire – Virginia

When advancing a hoseline into a fire structure, air should be bled from the line once it is charged, and before entering the structure.[4] Fire fighters should continually train in establishing a water supply, proper hose deployment, and advancing and operating hoselines to ensure successful interior attacks.

In this incident, after the 200 feet of 2 ½" attack line and 300 feet of 1 ¾" hoseline were deployed from E12, there were complaints of low water pressure in both lines. The officer from E12 removed some kinks from the 2 ½" attack line which had a positive effect on the pressure. The pressure at both the E12 and the supply pumper E20 supposedly reported no fluctuations. There may have been a combination of factors contributing to low water pressure, such as fluctuating residential water pressure, pressure problems at one or both engines, and/or supply line issues. There was insufficient information in this incident to make a determination as to the cause of low water pressure.

Recommendation #7: Fire departments should ensure that fire fighters are trained for extreme conditions such as high winds and rapid fire progression associated with lightweight construction

Discussion: Training is one of the most important steps in fire fighter safety. Fire fighters must strive to retain information and skills that are presented in training.[2] Training provides the necessary tools and fundamental knowledge to keep a fire fighter safe from injury. Just taking the training is not enough; the fire fighter needs to use their skills/information routinely on the fireground or at an actual emergency. Time is necessary to actually become proficient in those skills which are necessary for operational success in the field. In this era of new lightweight construction, training procedures covering strategy and tactics in extreme operational conditions, such as high winds and lightweight building construction (i.e., materials and design) are needed for all levels of fire fighters. Lightweight constructed buildings fail rapidly and with little warning, complicating rescue efforts.[11] The potential for fire fighters to become trapped or involved in a collapse may be increased. There are twenty-nine actions fire fighters can take to protect themselves when confronted with buildings utilizing lightweight building components as structural members. They range from looking for signs or indicators that these materials are used in buildings (such as, newer structures, large unsupported spans, and heavy black smoke being generated) to getting involved in newer building code development.[11]

Additionally,

Recommendation #8: Municipalities should ensure that dispatch collects and communicates information on occupancy and extreme environmental conditions

Discussion: The dispatch center should be aware of extreme environmental conditions on an hourly basis. Local weather forecasts and conditions are ready available in various media forms. When extreme weather conditions are present or the possibility exists, this information should be transmitted to the responding station when the call goes out. In addition, if the 911 caller does not relay any information about the occupancy of the fire structure, the dispatcher should explicitly ask the caller if they know if the structure is occupied.

Page 13

Fatality Assessment and Control Evaluation
Investigation Report # F2007-12

Career Fire Fighter Dies in Wind Driven Residential Structure Fire – Virginia

Although there is no evidence that the following recommendation could have specifically prevented this fatality, NIOSH investigators recommend that:

Recommendation #9: Fire departments ensure that radios are operable in the fireground environment

Discussion: The fireground communications process combines electronic communication equipment, a set of standard operating procedures, and the fire personnel who will use the equipment. To be effective, the communications network must integrate the equipment and procedures with the dynamic situation at the incident site, especially in terms of the environment and the human factors affecting its use. The ease of use and operation may well determine how consistently fire fighters monitor and report conditions and activities over the radio while fighting fires. Fire departments should review both operating procedures and human factors issues to determine the ease of use of radio equipment on the fireground to ensure that fire fighters consistently monitor radio transmissions from the IC and respond to radio calls.[12] The need to have properly functioning equipment during fire operations is critical.

In this incident, several fire fighters commented that radios were malfunctioning due to water shorting out the lapel microphone. However, the victim's radio was heard loud and clear during his mayday, along with several communications describing his believed location and request for water due to the extreme heat.

REFERENCES

1. Brunacini, A V [1985]. Fire Command. Quincy, MA: National Fire Protection Association.

2. Dunn V (1992). Safety and survival on the fireground. Saddlebrook, NJ: Fire Engineering Books and Videos.

3. USFA [2003]. Rapid Intervention Teams and How to Avoid Needing Them-SPECIAL REPORT. USFA–TR-123, March 2003.

4. International Fire Service Training Association [1998]. Essentials of Fire Fighting, 4th ed. Stillwater, Ok: Fire Protection Publications, Oklahoma State University.

5. NFPA (2004). NFPA 1710: Standard for the Organization and Deployment of Fire Suppression Operations, Emergency Medical Operations, and Special Operations to the Public by Career Fire Departments. Quincy, MA: National Fire Protection Association.

6. Angulo RA, Clark BA, Auch S [2004]. You Called Mayday! Now What? Fire Engineering, September issue.

Page 14

Fatality Assessment and Control Evaluation
Investigation Report # F2007-12

Career Fire Fighter Dies in Wind Driven Residential Structure Fire – Virginia

7. Clark BA [2004]. Calling a mayday: The drill. [http://www.firehouse.com]. Date accessed: November 2004.

8. DiBernardo JP [2003]. A missing firefighter: Give the mayday. Firehouse, November issue.

9. Sendelbach TE [2004]. Managing the fireground mayday: The critical link to firefighter survial. http://cms.firehouse.com/content/article/article.jsp?sectionId=10&id=10287. Date accessed: January 2008.

10. Miles J, Tobin J [2004]. Training notebook: Mayday and urgent messages. Fire Engineering, April issue.

11. Smith J (2002). Strategic and Tactical Considerations on the Fireground. Upper Saddle River, New Jersey: Prentice Hall; pages 185-190.

12. Fire Fighter's Handbook [2000]. Essentials of Fire Fighting and Emergency Response. New York: Delmar Publishers

INVESTIGATOR INFORMATION

This incident was investigated by Matt Bowyer, General Engineer, and Virginia Lutz, Safety and Occupational Health Specialist, with the Fire Fighter Fatality Investigation and Prevention Program, Division of Safety Research at NIOSH. Vance Kochenderfer, NIOSH Quality Assurance Specialist, National Personal Protective Technology Laboratory, conducted an evaluation of the victim's self-contained breathing apparatus. An expert technical review was conducted by Battalion Chief of Safety John J. Salka, Jr., New York City Fire Department.

Fatality Assessment and Control Evaluation
Investigation Report # F2007-12

Career Fire Fighter Dies in Wind Driven Residential Structure Fire – Virginia

APPENDIX

Summary of Status Investigation Report

NIOSH Task No. TN-15210

Background

As part of the ***National Institute for Occupational Safety and Health (NIOSH) Fire Fighter Fatality Investigation and Prevention Program***, the Technology Evaluation Branch agreed to examine and evaluate one Mine Safety Appliances 4500 psi self-contained breathing apparatus (SCBA).

This SCBA status investigation was assigned NIOSH Task Number TN-15210. The submitter was advised that NIOSH would provide a written report of the inspections and any applicable test results.

The SCBA, sealed in a corrugated cardboard box, was delivered to the NIOSH facility in Bruceton, Pennsylvania on May 25, 2007. Upon arrival, the sealed package was taken to the Firefighter SCBA Evaluation Lab (Building 108) and stored under lock until the time of the evaluation.

SCBA Inspection

The package was opened and the SCBA inspection was performed on June 20, 2007. The SCBA was inspected by Vance Kochenderfer, Quality Assurance Specialist, of the Technology Evaluation Branch, National Personal Protective Technology Laboratory (NPPTL), NIOSH. The SCBA was examined, component by component, in the condition as received to determine its conformance to the NIOSH-approved configuration. The entire inspection process was videotaped. The SCBA was identified as a Mine Safety Appliances (MSA) model; however, the damage was too extensive to determine the exact type.

The unit is extremely fire-damaged. Most of the plastic, rubber, and fabric components of the SCBA have been consumed. No performance testing could be conducted on the unit.

Personal Alert Safety System (PASS) Device

An ICM 2000 Plus Personal Alert Safety System (PASS) device was incorporated into the pneumatics of the SCBA. During the inspection, the PASS device could not be activated. The case was opened and representatives of MSA were able to retrieve stored data from the unit, and the last five uses are

Page 16

Fatality Assessment and Control Evaluation
Investigation Report # F2007-12

Career Fire Fighter Dies in Wind Driven Residential Structure Fire – Virginia

presented in **Appendix II** of the full Status Investigation Report. The data indicate that the unit's battery was exhausted six minutes into the last use while the cylinder pressure was 3350 psi and the internal temperature 130°F. From the limited data available there is no indication of unusual performance of the SCBA.

Summary and Conclusions

An SCBA was submitted to NIOSH for evaluation. The SCBA was delivered to NIOSH on May 25, 2007 and inspected on June 20, 2007. The unit was identified as an MSA 4500 psi SCBA, but the exact model and NIOSH approval number could not be determined. The SCBA has suffered severe fire damage and is not functional.

It is difficult to draw conclusions about the unit given its state. The cylinder valve was found to be fully open and the cylinder empty, which would be consistent with the SCBA being used to cylinder exhaustion. Data retrieved from the ICM 2000 Plus PASS device do not suggest any malfunction during the last recorded use.

In light of the information obtained during this investigation, the Institute has proposed no further action at this time. Following inspection and testing, the SCBA was returned to the package in which it was received and stored under lock in Building 108 at the NIOSH facility in Bruceton, Pennsylvania, pending return to the submitter.

Due to the extensive damage to the unit, it does not appear possible for it to be returned to service and it should be replaced.

Page 17

Appendix D
Tactical Worksheets

Temple Terrace Fire Department
Fireground Tactical Worksheet

Date: _____ Time: _____
Planning Section: _____
All Clear Time: _____

Address: _____
Safety Officer: _____
Under Control Time: _____

PIO: _____
Loss Stopped Time: _____

Strategy
☐ Offensive ☐ Marginal ☐ Defensive

Units	
Engine	
Engine	
Engine	
Engine	
Engine	
Ladder	
Ladder	
Rescue	
Rescue	
Chief	

Exposure D
Side D
Exposure C — | B | C | — Exposure A
        | A | D |
Side B — Side A
Exposure B

Incident Priorities
Life Safety
▲ ☐ Primary Search
▲ ☐ Evacuation
  ☐ Back-up Line
▲ ☐ Secondary Search
  ☐ Tactical Channel
  ☐ Rehab
Incident Stabilization
  ☐ Exposure
▲ ☐ Water on Fire
  ☐ Utility Disconnect
  ☐ Water Supply
  ☐ Sprinkler
  ☐ Ventilation
  ☐ Extension
  ☐ Lighting
  ☐ Air Supply
  ☐ CO Reading
  ☐ Crowd Control
▲ ☐ Under Control
▲ ☐ Loss Stopped
▲ ☐ Notify PIO
▲ ☐ Use of R-911
Property Conservation
  ☐ Salvage
  ☐ Overhaul
  ☐ Property Secured
  ☐ Fire Investigation
  ☐ Occ. Asst (RedCrs)

Accountability
▲ ☐ All Clear
▲ ☐ Under Control
▲ ☐ Loss Stopped
▲ ☐ 15 Minutes
▲ ☐ 30 Minutes
▲ ☐ 45 Minutes
▲ ☐ 1 Hour
▲ ☐ 1-¼ Hours
▲ ☐ 1-½ Hours
▲ ☐ 1-¾ Hours
▲ ☐ 2 Hours

Sector/Division/Groups					
Officer Assigned/Units					
Radio Channel					
Entry Time					

Structure Fire Tactical Worksheet

City of Santa Fe Fire Department

Location: _____ Time: _____ Date: _____

Tactical Benchmarks
- ☐ Safety
- ☐ Overall Plan
- ☐ Water Supply
- ☐ Survey
- ☐ Utilities
- ☐ Accountability
- ☐ Command Location
- ☐ Assignments
- ☐ Tactical Channels
- ☐ PD

Search & Rescue
- ☐ All Clear
- ☐ Primary Search
- ☐ Secondary Search

Ventilation
- ☐ Vertical
- ☐ Horizontal
- ☐ PPV

Exposures
- ☐ Side A
- ☐ Side B
- ☐ Side C
- ☐ Side D
- ☐ Above
- ☐ Below

Status
- ☐ Under Control
- ☐ Investigation
- ☐ Extinguishment

Command
- Liaison
- Safety
  - Rehab
- Public Information
- Operations

Resources

Engines

Tankers

Other

EMS

Rescue

Aerials

Brush Trucks

Scene Diagram

C

B   D

A

Elapsed Time	5	10	15	20	25	30	PAR

Structure Fire Tactical Worksheet

Glossary

Courtesy of Chief Chris Miekal, New Orleans Fire Department, LA.

Glossary

A

Ambient Conditions – Common, prevailing, and uncontrolled atmospheric weather conditions. The term may refer to the conditions inside or outside of the structure.

Autoignition Temperature (AIT) – The temperature to which the surface of a substance must be heated for ignition and self-sustained combustion to occur. The autoignition temperature of a substance is always higher than its piloted ignition temperature.

B

Backdraft – The explosive or rapid burning of heated gases that occurs when oxygen is introduced into a compartment that has a depleted supply of oxygen due to an existing fire.

Balloon Frame Construction – Construction method using long continuous studs that run from the sill plate (located on the foundation) to the roof eave line. All intermediate floor structures are attached to the studs. Requires the use of long lumber and generally lacks any type of fire stopping within the wall cavity.

Bearing Walls – Walls of a building that by design carry at least some part of the structural load of the building in the direction of the ground or base.

C

Collapse Zone – The area extending horizontally from the base of the wall to one and one-half times the height of the wall.

Compartment – Room or space within a building or structure which is enclosed on all sides, at the top, and bottom. The term *compartment fire* is defined as a fire that occurs within such a space.

Conduction – Physical flow or transfer of heat energy from one body to another through direct contact or an intervening medium from the point where the heat is produced to another location or from a region of high temperature to a region of low temperature.

Convection – Transfer of heat by the movement of heated fluids or gases; usually in an upward direction.

F

Fireground – Area around a fire and occupied by fire fighting forces.

Fireproof – Anything that is impervious to fire; also the act of protecting a structure or construction component from fire.

Fire Stop – Solid materials, such as wood blocks, used to prevent or limit the vertical and horizontal spread of fire and the products of combustion in hollow walls or floors, above false ceilings, in penetrations for plumbing or electrical installations, in penetrations of a fire-rated assembly, or in cocklofts and crawl spaces.

First Alarm Assignment – Initial fire department response to a report of an emergency; the assignment is determined by the local authority based on available resources, the type of occupancy, and the hazard to life and property.

Flashover – A transitional phase in the development of a compartment fire in which surfaces exposed to thermal radiation reach ignition temperature more or less simultaneously and fire spreads rapidly throughout the space resulting in full room involvement or total involvement of the compartment or enclosed area.

Freelancing – Operating independently of the IC's command and control.

Fuel – Flammable and combustible substances available for a fire to consume. Fuel can be in a solid, liquid, or gas form. In terms of a structure fire, the building, and its contents.

Fuel Load – The total quantity of combustible contents of a building, space, or fire area, including interior finish and trim, expressed in heat units or the equivalent weight in wood.

H

Heat – Form of energy associated with the motion of atoms or molecules in solids or liquids that is transferred from one body to another as a result of a temperature difference between the bodies such as from the sun to the earth. To signify its intensity, it is measured in degrees of temperature.

Heat Release Rate – The rate at which heat is generated by fire. The heat release rate is measured in Joules per second (also called Watts). Since a fire puts out much more than 1 Watt, the heat release rate is usually quantified in kilowatts (1000 W) or megawatts (a million watts).

K

Kerf Cut – The groove, cut, notch, or channel made by a saw or cutting tool.

L

Line-of-Duty Death (LODD) – Firefighter or emergency responder death resulting from the performance of fire department duties.

M

Mayday – International distress signal broadcast by voice.

O

Occupancy – (1) General term for a building, structure, or residence. (2) Building code classification based on the use to which owners or tenants put buildings or portions of buildings; regulated by the various building and fire codes. *See also* Occupancy Classification.

Occupancy Classification – Classifications given to structures by the model building code used in that jurisdiction based on the intended use for the structure.

P

Personnel Accountability Report (PAR) – A roll call of all units (crews, teams, groups, companies, sectors) assigned to an incident. Usually by radio, the supervisor of each unit reports the status of the personnel within the unit at that time. A PAR may be required by SOP at specific intervals during an incident, or may be requested at any time by the IC or the ISO.

Piloted Ignition – Occurs when a mixture of fuel and oxygen encounter an external heat source with sufficient heat energy to start the combustion reaction.

Platform Frame Construction – A construction method in which a floor assembly creates an individual platform that rests on the foundation. Wall assemblies the height of one story are placed on this platform and a second platform rests on top of the wall unit. Each platform creates fire stops at each floor level which restricts the spread of fire within the wall cavity.

Plenum – In a structure, the space between the real ceiling and a suspended ceiling. It is often used as a return air duct for the heating and air conditioning system. It may also contain electrical, telephone, and communication wires.

Preincident Plan – Plan developed from the information gathered during the preincident survey and used during emergency operations and training.

Preincident Survey – The act of collecting information on a site prior to the occurrence of an incident.

Protected Steel – Steel beams that are covered with either spray-on fire proofing (an insulating barrier) or fully encased in an Underwriters Laboratories Inc. (UL) designed system

Pyrolysis – Thermal or chemical decomposition of fuel because of heat that generally results in a lowered ignition temperature of the material. The preignition combustion phase of burning during which heat energy is absorbed by the fuel, which in turn gives off flammable tars, pitches, and vapors.

Q

Q-Deck – Also known as S-Deck and corrugated steel deck. Structural steel decks used to support concrete flooring and roof membranes. May have fire retardant material applied to the underside of the structure.

R

Radiation – The transmission or transfer of heat energy from one body to another body at a lower temperature through intervening space by electromagnetic waves such as infrared thermal waves, radio waves, or X rays. Also called Radiated Heat.

RECEO/VS Model – One of many models for prioritizing activities at an emergency incident: Rescue, Exposures, Confinement, Extinguishment, Overhaul, Ventilation, and Salvage.

Required Fire Flow – The estimated uninterrupted quantity of water expressed in gallons per minute (gpm) (liters per minute [L/min]) that is needed to extinguish a fire.

S

Search Line – Consists of 200 feet (60 m) of ⅜-inch (10 mm) rope with a Kevlar® sheath to provide abrasion protection and heat resistance. Steel rings are located at 20 foot intervals (6 m) for the attachment of tether or branch lines used for performing lateral searches from the main search line.

Situational Awareness – Being aware of your surroundings and what is going on.

Size-Up – Ongoing mental evaluation process performed by all firefighters on the scene. Size-up is the evaluation of visual clues and collection of information that decisions are based on. It is an ongoing process as initial decisions are continually checked with new information and fire indicators to confirm initial perception and changes to conditions as the fire operations progress. Size-up results in an action plan that may be adjusted as the situation changes. It includes such factors as time, location, nature of occupancy, life hazard, exposures, property involved, nature and extent of fire, weather, and fire fighting resources.

Structural Collapse – Structural failure of a building or any portion of it resulting from a fire, snow, wind, water, or damage from other forces.

T

Target Hazard – Facility or site in which there is a high potential for life or property loss.

Trenching – In strip or trench ventilation, the process of opening a roof area the width of the building with an opening 2 foot (0.6 m) wide to channel out fire and heat.

Two-In/Two-Out Rule – Occupational Safety and Health Administration (OSHA) regulation requiring a team of at least two personnel to be organized before entering an Immediately Dangerous to Life or Health (IDLH) atmosphere. It requires a standby team of at least two personnel outside the IDLH atmosphere to back up the entry team in the event they require rapid rescue.

Index

Courtesy of Ron Moore.

Index

A

Abandoned occupancy, life hazard, 127–128
Access
    access points, 116, 120
    access time, 31, 32
    barriers to access, 116
    egress, 122–123
    metal security bars, 123
    street access, 122
    structure access, 122
Accountability, personnel
    check-in, 21
    Incident Action Plan, 21
    personnel accountability report, 147, 152
    purpose of, 20
    resource tracking, 21
    span of control, 21
    systems, 20
    tracking of personnel, 150, 152
    Two-In/Two-Out Rule, 35, 110
    Unity of Command, 21
Aerial devices, 116, 130
AHJ. *See* Authority having jurisdiction (AHJ)
Air conditioning. *See* Heating, ventilating, and air conditioning (HVAC)
Air management, life safety hazards, 124
Air track of smoke, 135
AIT (autoigniton temperature), 48
Alarm systems, 129
Alterations, size-up factors, 123. *See also* Renovations
Ambient conditions, 59
American Society for Testing and Materials (ASTM), time-temperature curve, 79
Apparatus, size-up factors, 128
Arched roofs, 119
Area of floor space, 116, 120
Arrival at the incident
    arrival report, 107–108
    size-up upon arrival, 106–110, 131–136
    visual size-up, 108
The Art of War, 43
Assembly occupancy, life hazards, 126
ASTM (American Society for Testing and Materials), time-temperature curve, 79
Attacking and Extinguishing Interior Fires, 161
Attics
    building construction, 75–76
    concealed horizontal spaces, 61
    size-up factors, 121
    storage as collapse factor, 80
    visual indicators, 132
Attitude of personnel, 33

Authority having jurisdiction (AHJ)
    building codes, 63
    cause and origin of fire, 175
    responsibilities, 9
Autoignition, 53
Autoignition temperature (AIT), 48
Automatic aid, 32–33

B

Backdraft
    conditions for, 51–52
    defined, 51
    indicators, 52
    variable factors, 52
Balloon frame construction, 60
Barriers to access or egress, 116
Barriers to aerial operations, 116
Basements, 73–75, 132
Bearing walls, 67
Big Fire, Big Water, 101
Black fire, 134
Bowstring roof, 119
Brace frame building construction, 60
Branch, defined, 13
Breathing air supply apparatus, 130
Building characteristics during size-up, 115–123
    accessibility, 122–123
    building styles, 117
    exterior view, 116
    hazardous conditions, 123
    preincident plan, 115–116
    roof types, 118–120
    size and configuration, 120–121
    types of construction, 117–118
Building codes
    early codes, 70–71
    existing building exemptions, 60
    IBC®, 63–69
    ICC, 63
    local codes, 63
    manufactured (mobile) homes, 63, 68
    NBC, 69
    NFPA®, 63–69
    purpose of, 60
Building construction
    attics and cocklofts, 75–76
    balloon frame construction, 60
    basements, 73–75
    brace frame, 60
    building codes. *See* Building codes
    building configuration, 56

Building construction *(continued)*
    building periods or age, 70–72
        early regulatory codes, 70–71
        new material uses, 71–72
        suburban development, 71
    Canadian construction, 69–70
    cellar, 73–74
    collapse potential, 76–82
    combustible interior finishes, 57
    concealed spaces and voids, 61–62, 76
    condominiums, 61
    contents as fuel, 57
    crawl spaces, 73–74
    early regulatory codes, 70–71
    existing building code exemptions, 60
    interior arrangement, 72–73
        compartmentalization, 73
        fire behavior and, 61
        open floor plans, 72–73
    material fuel load, 56
    McMansions, 61
    mortise and tenon, 60
    occupancy classifications, 43
    petroleum-based materials, 55
    platform frame construction, 60
    post and beam, 60
    preincident planning, 36, 104–105
    projections during size-up, 105–106
    rain roof, 62, 120. *See also* Roofs
    structural collapse, 62
    town houses, 61
    types, 63
    types and materials for the area, 60
    United States construction, 63–69
        Type I (fire resistive), 64
        Type II (noncombustible or limited combustible), 65
        Type III (ordinary construction), 65–66
        Type IV (heavy timber/mill construction), 66
        Type V (wood frame), 67–68
        unclassified, 68–69
    wood and plywood, 56, 71
Building contents
    contents as fuel, 57
    fire ignition in, 88
    fire intensity, 88
    fuel load, 88
    pyrolysis, 57
    structural collapse factor, 80
Building periods or age, 70–72
    early regulatory codes, 70–71
    new material uses, 71–72
    suburban development, 71
Business occupancy, life hazards, 126

C

CAAN (conditions, actions, air, needs) 111
CAF (compressed air foam) units, 130
CAN (conditions, actions, needs) 111
Canadian construction, 69–70

CARA (conditions, actions, resources, air) 111
Cardiac arrest deaths, 1
Ceiling height
    fire development, 58
    size-up factors, 116, 120
Cellars, 73–74
Check-in, 21
Chief, defined, 14
Class A foam, 174–175
Class A or cellulose type fuel, 54, 55
Class B foam, 175
Class B type fuel, 54, 55
Class C type fuel, 54
Class D type fuel, 54
Class K type fuel, 54
Clear text communications, 16
Cm-deck roof, defined, 118
Cockloft space
    building construction, 75–76
    concealed horizontal spaces, 61
    size-up factors, 121
    storage as collapse factor, 80
    visual indicators, 132
Cocoanut Grove fire, 57
Code of Federal Regulations–Title 24: Housing and Urban Development, 68
Collapse. *See* Structural collapse
Collapse zone, 78–79
Combustible construction, 69
Command, defined, 12
Command options
    Command option (command post option), 153–154
    Command Post option, 153–154
    fast-attack or mobile command option, 153
    investigation option (nothing-showing), 153
    transferring Command, 154
Command Post option, 153–154
Command staff, defined, 12
Communications
    cellular telephones, 30
    clear text, 16
    common communications, 15–16
    face-to-face, 149–150
    Incident Action Plan, 149
    landline telephones, 30
    pagers, 30
    radios
        capabilities, 16
        frequencies or channels, 16
        resource tracking, 149–150
        resources, 30
    resource tracking, 149–150
Compartment
    ambient conditions, 59
    ceiling height and fire development, 58
    defined, 45
    fire behavior, 47–48
        configurations, 44
        mushrooming, 47

thermal layering, 47
fire spread, 45
thermal properties, 59
ventilation and fire development, 58
volume and fire development, 57
Compartmentalization interior building arrangement, 73
Comprehensive resource management, 20
Compressed air foam (CAF) units, 130
Concealed space
building construction, 61–62, 76
functions, 76
size-up factors, 121
Condominiums, 61
Conduction, defined, 45
Confinement of the fire
defined, 171
factors determining tactics, 171–172
before rescue operations, 162
size-up, 172
tasks, 173
trenching, 172–173
ventilation, 172
Construction. See Building construction
Contents. See Building contents
Convection, defined, 45
Crawl spaces, 73–74
Crew
defined, 14
resources, 125
Curtain collapse, 78

D

Day of week and response considerations, 105
Dead-end corridors, 123
Decision-making
examples, 24
facts, 104–105
identify the problem, 25
indecision, 26
model, 24
monitoring results, 26
perceptions, 105
process steps, 24
projections, 105–106
size-up considerations, 112, 114–115
solution, determination of, 25–26
solution implementation, 26
Defensive strategy
applicability, 145–146
risk outweighs benefit, 146
structural condition, 146
structural deterioration, 146
volume of fire, 145
causes of fatalities and injuries, 146–147
line-of-duty death, Prince William County, 147–148
personnel accountability reports, 147
purpose of, 23
transition from offensive strategy, 147

Department of Housing and Urban Development (HUD), 63, 68
Detection systems, 129
Director, defined, 14
Dispatch broadcast information, 102–103
Dispatch time, 31
Division, defined, 13
Doors as means of egress, 122
Dwelling. See Residence

E

Earthquakes, 79
Educational occupancy, life hazards, 126–127
Egress, means of, 122–123
Electrical wiring, 71
Exposures
building characteristics, 116
defined, 168
exterior, 168
interior, 168
Exterior exposures, 169–171
Extinguishment, 173–175
Class A foam, 174–175
Class B foam, 175

F

Facilities at incidents
Incident Command Post, 20
staging areas, 20
titles and functions, 18
Factory Mutual (FM) formula, 100
Factory-built homes. See Manufactured (mobile) home
Facts
construction and occupancy, 104–105
defined, 104, 112
fire behavior, 105
overhaul, 176
resources, 105
site knowledge, 104
size-up considerations, 112–113
time of day/day of week, 105
Fast-attack or mobile command option, 153
Fire behavior
in compartments, 47–48
configurations, 44
mushrooming, 47
thermal layering, 47
factors affecting fire development, 54–59
additional fuel, 56–57
ambient conditions, 59
combustible interior finishes, 57
compartment volume and ceiling height, 57–58
flashover transition, 50
fuel load, 59
fuel type, 54–55
thermal properties of the compartment, 59
ventilation, 58

Index **395**

Fire behavior *(continued)*
    fire growth theories, 44
    fire spread, 45
    importance of knowledge of, 43
    preincident planning, 105
    projections, 105
    rapid fire development, 48–53
        autoignition, 53
        backdraft, 51–52
        conditions, 48
        flashover, 49–50
        indicators, 49
        rollover, 52–53
        smoke explosion, 53
Fire development
    additional fuel, 56–57
    ambient conditions, 59
    combustible interior finishes, 57
    compartment volume and ceiling height, 57–58
    flashover transition, 50
    fuel load, 59
    fuel type, 54–55
    thermal properties of the compartment, 59
    ventilation, 58
Fire escapes, 123
Fire Fighting Tactics, 112, 161
Fire load, 59
Fire protection systems
    detection and alarm systems, 129
    generally, 129
    non-water-based, 129
    size-up factors, 116, 128–129
    sprinklers, 60
    standpipe system, 129
    water-based, 129
Fire spread, 45
Fire stop, defined, 65
Firefighter Level I, knowledge of fire behavior, 43
Fireground
    defined, 142
    life safety, 142
    projections during size-up, 106
Fireproof, defined, 64
First alarm assignment, 25
Flame color, structural fire condition indicators, 136
Flashover
    conditions for, 50
    defined, 49
    determining factors, 50
    elements
        compartment, 50
        ignition of all exposed surfaces, 50
        rapidity, 50
        transition in fire development, 50
    rapid fire development, 49–50
Flat roofs, 118
Floors
    above grade, 121
    area of floor space, 116, 120
    below grade, 120–121
    open floor plans, 72–73
FM (Factory Mutual) formula, 100
Freelancing, 26, 149
Fuel
    additional fuel for the fire, 56–57
    defined, 45
    fire development factors, 54–55
    fuel load
        building contents, 88
        defined, 43, 59
    types
        Class A or cellulose, 54, 55
        Class B, 54, 55
        Class C, 54
        Class D, 54
        Class K, 54
        gas leak, 55
Fundamentals of Fire Fighting Tactics, 161

G

Gases
    gas leaks as fuel, 55
    pressure, 47–48
    thermal layering, 47
General staff, defined, 13
Group, defined, 13

H

Hazards, preincident planning, 38
Heat
    defined, 45
    reflectivity, 59
    structural fire condition indicators, 136
    transfer, 45
        conduction, 45
        convection, 45
        radiation, 45
Heating, ventilating, and air conditioning (HVAC)
    concealed vertical spaces, 61
    early regulatory codes, 71
    fire development factors, 58
Heavy timber construction, 69
Hierarchy of command decisions, 21–22
Homes. *See* Residence
Hose tenders, 130
Hoselines for firefighter protection, 167, 180–181
HUD (Department of Housing and Development), 63, 68
Humidity, impact on operations, 131
HVAC. *See* Heating, ventilating, and air conditioning (HVAC)
Hybrid modular home, 69

I

IAFC (International Association of Fire Chiefs), *The 10 Rules of Engagement for Structural Fire Fighting*, 143
IAP. *See* Incident Action Plan (IAP); Initial Action Plan (IAP)
IBC®. *See* International Building Code® (IBC®)
IC. *See* Incident Commander (IC)

ICC. *See* International Code Council (ICC)
Ice, impact on operations, 131
ICP (Incident Command Post), 20
ICS. *See* Incident Command System (ICS)
Identify the problem, 25
IDLH. *See* Immediately dangerous to life and health (IDLH)
Immediately dangerous to life and health (IDLH)
    air management, 124
    fast-attack or mobile command option, 153
    personnel accountability, 152
    resource capabilities, 103
    search of IDLH area, 166
    two-in/two-out rule, 110
    victim removal, 167–168
IMS (Incident Management System), 10
Incident Action Plan (IAP)
    on arrival at the scene, 109–110
    defined, 17
    incident facilities, 18, 20
    personnel accountability, 21
    purpose of, 18
    for scene management
        Command option (Command Post option), 153–154
        Command options, 153–154
        fast-attack or mobile Command option, 153
        investigation option (nothing-showing), 153
        plan development, 148–149
        plan implementation, 152–155
        resource allocation, 154–155
        transferring Command, 154
    tactical work sheets, 18, 150
    verbal plan, 18
Incident Command Post (ICP), 20
Incident Command System (ICS)
    background, 10–11
    characteristics, 11–12
    common communications, 15–16
    comprehensive resource management, 20
    facilities, 18, 20
    Incident Action Plan, 17–18. *See also* Incident Action Plan (IAP)
    modular design, 15
    personnel accountability, 20–21
    purpose of, 10–11
    scalable design, 15
    span of control, 18
    terminology
        leadership titles, 14–15
        organizational levels, 12–13
        resources, 14
    unified command structure, 16–17
    Unity of Command, 17
Incident Commander (IC)
    defensive tactics, 146–147
    defined, 14
    knowledge of fire behavior, 43
    offensive strategy, 145
    personnel accountability, 150, 152
    personnel accountability reports, 147
    resource allocation, 154–155
    resource tracking, 149
    responsibilities, 9, 11
    skills needed, 34
    Tactical Work Sheet, 150
Incident Management System (IMS), 10
Incident priorities
    hierarchy of command decisions, 21–22
    incident stabilization, 22, 142
    life safety, 22, 142
    property conservation, 22, 142
    strategy, 141–142
Incident Safety Officer (ISO), defined, 12
Incident stabilization
    defined, 22
    incident priorities, 22
    strategy, 142
Industrial occupancy, life hazards, 126
Initial Action Plan (IAP), 17
Institutional occupancy, life hazards, 126
Insulation, thermal properties, 59
Insurance Services Office (ISO), 94, 100
Interior arrangement, 72–73
    compartmentalization, 73
    fire behavior and, 61
    open floor plans, 72–73
Interior building arrangement, 72–73
    compartmentalization, 73
    fire behavior and, 61
    open floor plans, 72–73
Interior exposures, 169, 171
International Association of Fire Chiefs (IAFC), *The 10 Rules of Engagement for Structural Fire Fighting*, 143
International Building Code® (IBC®)
    occupancy classifications, 83–88
    types of construction, 63–69
International Code Council (ICC)
    fire flow calculations, 100
    types of construction, 63
Investigation option (nothing-showing), 153
ISO (Incident Safety Officer), defined, 12
ISO (Insurance Services Office), 94, 100

K

Kerf cut, defined, 172
Key information, 3–4
Knowledge of personnel
    examples, 34
    experiential, 34
    factual, 34

L

Layman, Lloyd
    Attacking and Extinguishing Interior Fires, 161
    decision-making, 114–115
    Fire Fighting Tactics, 112, 161
    Fundamentals of Fire Fighting Tactics, 161
    overhaul, 175

RECEO/VS model, 155–156, 161. *See also* RECEO/VS model
size-up process, 17–18
Leader, defined, 14
Leadership titles
    chief, 14
    director, 14
    Incident Commander, 14. *See also* Incident Commander (IC)
    leader, 14
    manager, 14
    officer, 15
    supervisor, 15
Life hazard, size-up, 124–128
    air management, 124
    assembly, 126
    crew resources, 125
    educational, 126–127
    general considerations, 124–125
    industrial, 126
    institutional, 126
    mercantile and business, 126
    rehabilitation, 125
    residential, 125
    unoccupied, vacant, abandoned structures, 127–128
Life safety
    defined, 22
    incident priorities, 22, 142
    strategy, 142
    your world, 27
Line-of-duty-deaths (LODDs)
    cardiac arrest as cause, 1
    defined, 143
    The 10 Rules of Engagement for Structural Fire Fighting, 143
LIP (Life Safety, Incident Stabilization, Property Conservation), 22–23, 141–142
LODD. *See* Line-of-duty-deaths (LODDs)

M
Manager, defined, 14
Managing the incident. *See also* Scene management
    authority having jurisdiction, responsibilities, 9
    decision-making, 24–26
        indecision, 26
        monitoring results, 26
        problem identification, 25
        solution implementation, 26
        solutions, determination of, 25–26
    Incident Action Plan, 148–149, 153–155
    Incident Command System
        background, 10–11
        characteristics, 11–12
        common communications, 15–16
        comprehensive resource management, 20
        facilities, 18, 20
        Incident Action Plan, 17–18
        modular and scalable, 15
        personnel accountability, 20–21
        purpose of, 10–11
        span of control, 18
        terminology, 12–15
        unified command structure, 16–17
        Unity of Command, 17
    Incident Commander, 9, 11. *See also* Incident Commander (IC)
    incident priorities, 22–23
    NIMS/ICS
        incident facilities, 18, 20
        overview, 9
        personnel accountability, 20–21
        purpose of, 10–11
    overview, 9–10
    preincident planning, 35–38
    resources, 26, 29–33
        automatic aid, 32–33
        mutual aid, 33
        outside aid, 33
        reflex time, 30–32
    strategies, 23
    tactics, 23
    tasks, 24
    terminology
        leadership titles, 14–15
        organizational levels, 12–13
        resources, 14
    your world, 26–29
    yourself, characteristics of, 33–35
        strengths, 33–35
        weaknesses, 35
Manufactured (mobile) home
    building codes, 63, 68
    description, 67, 68
Mayday, defined, 146
McMansions, 61
Means of egress, 122–123
Mercantile occupancy, life hazards, 126
Metal security bars, 123
MGM Grand Hotel fire, 57, 105
Mobile command option, 153
Mobile homes. *See* Manufactured (mobile) home
Modular design of ICS, 15
Modular home, 69
Monitoring results of decisions, 26
Mortise and tenon building construction, 60
Multiple use structure, 83, 88
Mushrooming effect, 47
Mutual aid, 33

N
National Building Code of Canada (NBC)
    occupancy classifications, 83–88
    types of building construction, 69
National Fire Academy (NFA), fire flow calculations, 100–101
National Fire Protection Association® (NFPA®). *See also specific NFPA®*
    building codes, 63–69
    cardiac arrest deaths, 1
    Fire Fighting Tactics, 161

Initial Action Plan, 17
injury types, 1
manufactured homes, 68
occupancy classifications, 83–88
RIC/RIT requirements, 106
vacant structure fires, 127
National Incident Management System/Incident Command System (NIMS/ICS)
incident facilities, 18, 20
overview, 9
personnel accountability, 20–21
purpose of, 10–11
National Institute of Standards and Technology (NIST), potential structural collapse, 81
NBC. *See* National Building Code of Canada (NBC)
Neutral plane, 48, 134–135
NFA (National Fire Academy), fire flow calculations, 100–101
NFPA®. *See* National Fire Protection Association® (NFPA®)
NFPA® 1021, *Standard for Fire Officer Professional Qualifications*, 17, 175
NFPA® 1500, *Standard on Fire Department Occupational Safety and Health Program*, 94, 152
NFPA® 1561, *Standard on Emergency Services Incident Management System*, 18
NFPA® 1620, *Standard for Pre-Incident Planning*, 37, 95
NFPA® 1710, *Standard for the Organization and Deployment of Fire Suppression Operations, Emergency Medical Operations, and Special Operations to the Public by Career Fire Departments*, 94
NFPA® 1720, *Standard for the Organization and Deployment of Fire Suppression Operations, Emergency Medical Operations and Special Operations to the Public by Volunteer Fire Departments*, 94
NIMS/ICS. *See* National Incident Management System/Incident Command System (NIMS/ICS)
NIST (National Institute of Standards and Technology), potential structural collapse, 81
Noncombustible construction, 69
Non-water-based fire protection system, 129
Nothing-showing command option, 153
Novoclimat standard, 69–70
Number of stories, 116

O

Occupancy classification
assembly, 126
business, 126
changes, 123
defined, 43
educational, 126–127
IBC®, 83–88
industrial, 126
institutional, 126
mercantile, 126
multiple use, 83, 88
NBC, 83–88
NFPA®, 83–88
preincident planning, 104–105
preincident survey, 94, 96
single use, 83
size-up factors, 116, 123
unoccupied, vacant, abandoned, 127–128
Occupational Safety and Health Administration (OSHA)
RIC/RIT requirements, 106
two-in/two-out rule, 110
Offensive strategy, 23, 145
Officer, defined, 15
Open floor plans, 72–73
Operational strategies, 144–148
defensive strategy, 145–147
line-of-duty death, Prince William County, 147–148
offensive strategy, 145
Organization of this manual, 2–3
OSHA. *See* Occupational Safety and Health Administration (OSHA)
Outside aid, 33
Overhaul, 175–178
cause of fire, 175
defined, 175
facts before performing, 176
liability of fire department, 175
place of operations, 176–177
risks, 177
tasks, 175, 178
Own situation, size-up considerations, 112, 114

P

Pagers, 30
Panelized home, 69
PAR (personnel accountability report), 147, 152
Perceptions during size-up, 105
Personnel
accountability
check-in, 21
Incident Action Plan, 21
personnel accountability report, 147, 152
purpose of, 20
resource tracking, 21
span of control, 21
systems, 20
tracking of personnel, 150, 152
two-in/two-out rule, 35, 110
unity of command, 21
allocation at the incident scene, 155
resources, 29
size-up factors, 130
strengths, 33–35
attitude, 33
knowledge, 34
skills, 34–35
weaknesses, 35
Phoenix, Arizona, Fire Department SOP model, 144
Piloted ignition, defined, 48
Pipe chases, 61
Pitched roofs, 119
Plan of operation
defined, 17–18
size-up considerations, 115

Platform frame construction, defined, 60
Plenum, defined, 76
Plywood
    as fire fuel, 56
    roof sheathing, 71
Point of no return, 124
Post and beam building construction, 60
POV (privately owned vehicle), 32, 152
Pre-cut home, 69
Preincident plan
    building characteristics, 115–116
    building construction, 36, 104–105
    characteristics, 37–38
        accessible, 37
        accurate, 38
        current, 38
    construction and renovation surveys, 36
    data analysis, 38
    defined, 94
    fire behavior and, 105
    functions, 36
    information gathering, 35
    NFPA® standards, 37
    process, 35
    purpose of, 35, 141
    reasons for, 36
    salvage, 181–182
    site plan, 38
    special hazards, 38
    target hazard, 36
Preincident survey
    before an alarm, 101–102
    at construction sites, 96–97
    defined, 94
    exposure information, 168–169
    frequency of review, 96
    information gained, 94
    information gathering, 97
    occupancies to survey, 94, 96
    preparation, 95
    process, 97–100
    reasons for, 95
    required fire flow calculations, 100–101
    standards, 94, 95
    target hazards, 96
    types of structures, 94
Pressure of gases, 47–48
Primary search, 165, 166–167
Priorities. See Incident priorities
Privately owned vehicle (POV), 32, 152
Probabilities, size-up considerations, 112, 113–114
Projections during size-up
    building construction, 105–106
    fire behavior, 105
    fire fighting activities, 106
Property conservation
    incident priorities, 22
    salvage, 181–183
    strategy, 142

Protected steel, defined, 64
Purpose of this manual, 2
Pyrolysis
    building contents, 57
    compartment fire development, 48
    defined, 45
    fuels in upper levels of buildings, 57
    plywood, 56
Pyrotechnics as fire ignition, 57

Q
Q-deck roof, defined, 118

R
Radiation, defined, 45
Radios
    capabilities, 16
    clear text communications, 16
    frequencies or channels, 16
    resource tracking, 149–150
Rain, impact on operations, 131
Rain roofs, 62, 120
Rapid fire development, 48–53
    autoignition, 53
    backdraft, 51–52
    conditions, 48
    flashover, 49–50
    indicators, 49
    rollover, 52–53
    smoke explosion, 53
Rapid intervention crew/team (RIC/RIT), 106
RECEO/VS model.
    background, 155
    confinement, 171–173
    defined, 155
    described, 156
    exposures, 168–171
        exterior exposures, 168, 169–171
        external tasks, 171
        interior exposures, 169, 171
        internal tasks, 171
    extinguishment, 173–175
    overhaul, 175–178
    salvage, 181–183
    search and rescue, 161–168
        conducting a search, 165–168
        safety guidelines, 164–165
    ventilation, 178–181
Reflex time
    access time, 31, 32
    defined, 30
    dispatch time, 31
    response time, 31, 32
    setup time, 31, 32
    turnout time, 31
    variables, 32
Rehabilitation, life safety hazards, 125
Rekindling of fire, 175, 177

Renovations
    alterations, size-up factors, 123
    preincident planning, 36
    size-up factors, 36
Required fire flow
    calculations, 100–101
    defined, 100
Rescue
    defined, 161, 162
    determination of need for, 162
    RECEO/VS model, 161–168
    tasks, 163
Residence
    building construction, 60–61
    condominiums, 61
    hybrid modular home, 69
    life hazard, 125
    manufactured (mobile) homes, 63, 67–68
    modular home, 69
    multistory McMansions, 61
    Novoclimat standard, 69–70
    open floor plans, 72–73
    panelized home, 69
    pre-cut home, 69
    salvage operations, 181–182
    SOP/SOGs for strategies and tactics, 94
    360-degree survey, 109, 132
    town houses, 61, 74–75
    volunteer firefighter fire deaths, 121
Resources
    agreements with others, 32
    allocation at the incident scene, 154–155
    apparatus, 29
    automatic aid, 32–33
    communication systems, 30
    comprehensive resource management, 20
    crew, 14, 125
    defined, 14
    incident management, 26, 29–33
    mutual aid, 33
    outside aid, 33
    personnel, 29
    preincident planning, 105
    reflex time, 30–32
    response considerations, 103
    single resources, 14
    size-up factors, 128–130
        apparatus, 128
        fire protection systems, 128–129
        other available resources, 130, 155
        water supply, 129–130
    strike team, 14
    task force, 14
    tracking, 21, 149–152
    water supply, 30
Response time
    defined, 31
    variables, 32
Retention, thermal properties, 59

RIC/RIT (rapid intervention crew/team), 106
Risk vs. benefit of strategy, 143–144, 174
Rollover
    conditions for, 52–53
    defined, 52
Romex® electrical wiring, 71
Roofs
    arched (bowstring), 119
    building characteristics, 116
    cm-deck, 118
    collapse, 62
    flat, 118
    generally, 118
    pitched, 119
    plywood sheathing, 71
    Q-deck, 118
    rain roof, 62, 120
    S-deck, 118
    size-up factors, 118–120

S
Safety Data Sheet (SDS), preincident planning, 38
Salvage, 181–183
    defined, 181
    preincident plan and size-up, 181–182
    residential properties, 181–182
    tasks, 183
Scalable design of ICS, 15
Scene management, 148–155. *See also* Managing the incident
    Incident Action Plan development, 148–149
    Incident Action Plan implementation, 152–155
    resource tracking, 149–152
Scope of this manual, 2
S-deck roof, defined, 118
SDS (Safety Data Sheet) preincident planning, 38
Search
    defined, 161, 162
    determination of need for, 162
    hoseline protection, 167
    primary search, 165, 166–167
    safety guidelines, 164–165
    search line, defined, 165
    secondary search, 166, 167
Secondary search, 166, 167
Section, defined, 13
Security bars, 123
Setup time, 31, 32
Sheltering-in-place, 161
Single resources, defined, 14
Single use structure, 83
Site plan, preincident planning, 38, 104
Situational awareness, 113
Size-up
    application, 93–111
        on arrival, 106–110
        facts, 104–105
        during the incident, 111
        perceptions, 105

Size-up *(continued)*
- preincident, 94–102
- projections, 105–106
- while responding, 102–104
- arrival condition indicators, 106–110, 131–136
  - structural fire condition indicators, 133–136
  - time of day, 131
  - visual indicators, 132
  - weather, 131–132
- building characteristics, 115–123
  - accessibility, 122–123
  - building styles, 117
  - exterior view, 116
  - hazardous conditions, 123
  - preincident plan, 115–116
  - roof types, 118–120
  - size and configuration, 120–121
  - types of construction, 117–118
- confinement, 172
- considerations, 111–115
  - decision, 112, 114–115
  - facts, 112–113
  - own situation, 112, 114
  - plan of operation, 112, 115
  - probabilities, 112, 113–114
- defined, 93
- extinguishment methods, 174
- hazardous conditions
  - alterations, 123
  - occupancy change, 123
  - storage, 123
- life hazard, 124–128
  - air management, 124
  - assembly, 126
  - crew resources, 125
  - educational, 126–127
  - general considerations, 124–125
  - industrial, 126
  - institutional, 126
  - mercantile and business, 126
  - rehabilitation, 125
  - residential, 125
  - unoccupied, vacant, abandoned structures, 127–128
- plan of operation, 17–18
- preincident, 94–102
  - before an alarm, 101–102
  - plans, 94
  - required fire flow calculations, 100–101
  - survey, 94–97
  - survey process, 97–100
- resources, 128–130, 155
- salvage, 181
- ventilation, 180
- visual size-up, 108

Skills
- defined, 34
- Incident Commander, 34
- purpose of, 34–35

Smoke
- air track, 135
- black smoke, 134
- black smoke from petroleum-based materials, 55, 132
- brown smoke, 133
- density, 135
- gray smoke, 134
- movement, 135–136
- smoke explosion, 53
- structural fire condition indicators, 133–136
- volume, 135
- white smoke, 133

Snow, impact on operations, 131

Solution
- determination of, 25–26
- implementation, 26

SOP/SOG. *See* Standard operating procedures/guidelines (SOP/SOG)

Span of control
- personnel accountability, 21
- purpose of, 18

Sprinklers, building codes, 60. *See also* Fire protection systems

Staging areas, 20

Stairs as means of egress, 123

Standard operating procedures/guidelines (SOP/SOG)
- dispatch broadcast information, 102–103
- Phoenix, Arizona, Fire Department model, 144
- preincident surveys, 94, 96
- residence strategies and tactics, 94
- resource tracking, 149
- size-up on arrival, 107, 108

Standpipe fire protection system, 129

Station nightclub fire, 57

Statistics
- annual emergency responses, 1
- injury types, 1
- types of incidents in 2008, 10

Storage, size-up factors, 123

Strategy
- defensive strategy, 23, 145–147
- hierarchy of command decisions, 21–22
- Incident Action Plan
  - command options, 153–154
  - development, 148–149
  - implementation, 152–155
  - resource allocation, 154–155
- incident priorities, 141–142
  - incident stabilization, 142
  - life safety, 142
  - property conservation, 142
- line-of-duty death, Prince William County, 147–148
- offensive strategy, 23, 145
- operational strategy, 144–148
- RECEO/VS model, 155–156
- risk vs. benefit, 143–144, 174
- scene management, 148–155
  - Incident Action Plan development, 148–149
  - Incident Action Plan implementation, 152–155

resource tracking, 149–152
Strike team, defined, 14
Structural collapse
    collapse zone
        defined, 78
        location of defensive operations, 79
        purpose of, 79
        stage of fire, 79
        time-temperature curve, 79
        Type I (fire resistive) construction, 78
        Type II (noncombustible or limited combustible) construction, 78
        Type III (ordinary construction), 78
        Type IV (heavy timber/mill construction), 78
        Type V (wood frame) construction, 78
    contents as factor, 80
    curtain collapse, 78
    defined, 76
    factors in determining potential for, 77–78
    indicators, 81
    lightweight construction, 81
    potential for, 62, 76–82
    preventing injury or death, 82
    roofs, 62
    during salvage operations, 182
    Vendome Hotel fire, 77
    water as factor, 80
Structure of buildings, information needed for incident management, 27
Suburban development of construction, 71
Sun Tzu, 43
Supervisor, defined, 15
Survey. *See* Preincident survey

T
Tactical work sheets, 18, 150
Tactics
    confinement, 171–173
    defined, 23
    examples, 23
    exposures, 168–171
        exterior exposures, 169–171
        external tasks, 171
        interior exposures, 169, 171
        internal tasks, 171
    extinguishment, 173–175
    hierarchy of command decisions, 21–22
    objectives, 23
    overhaul, 175–178
    salvage, 181–183
    search and rescue, 161–168
        conducting a search, 165–168
        safety guidelines, 164–165
    ventilation, 178–181
Target hazard, 36, 96
Task force, defined, 14
Tasks
    defined, 24
    examples, 24

hierarchy of command decisions, 21–22
Telephone communications, 30
Temperature extremes, impact on operations, 131–132
The 10 Rules of Engagement for Structural Fire Fighting, 143
Terminology
    leadership titles, 14–15
    organizational levels, 12–13
    resources, 14
Thermal Imaging Camera (TIC), 166
Thermal layering of gases, 47
Thermal properties of compartments, 59
360-degree survey, 109, 132
TIC (Thermal Imaging Camera), 166
Time considerations
    reflex time
        access time, 31, 32
        defined, 30
        dispatch time, 31
        response time, 31, 32
        setup time, 31, 32
        turnout time, 31, 32
        variables, 32
    time of alarm and response considerations, 103, 105
    time of day, arrival condition indicators, 131
    time of day/day of week, 105, 131
    time-temperature curve, 79
Topography, 116
Town houses
    basement fire, 74–75
    building construction, 61
Trenching, 172–173
Turnout time, 31, 32
Two-in/two-out rule
    on arrival at the scene, 110
    defined, 35
    OSHA requirements, 110
Type I (fire resistive) construction
    collapse zone, 78
    description, 64
    fuel for a fire, 88
    size-up factors, 117
Type II (noncombustible or limited combustible) construction
    collapse zone, 78
    description, 65
    fuel for a fire, 88
    size-up factors, 117
Type III (ordinary construction)
    collapse zone, 78
    description, 65–66
    fuel for a fire, 88
    size-up factors, 117–118
Type IV (heavy timber/mill construction)
    collapse zone, 78
    description, 66
    fuel for a fire, 88
    size-up factors, 118
Type V (wood frame) construction
    collapse zone, 78

Type V (wood frame) construction *(continued)*
    description, 67–68
    fuel for a fire, 88
    size-up factors, 118

U
Unclassified construction, 68–69
Unified Command Structure, 16–17
Unit, defined, 13
United States construction, 63–69
    Type I (fire resistive), 64
    Type II (noncombustible or limited combustible), 65
    Type III (ordinary construction), 65–66
    Type IV (heavy timber/mill construction), 66
    Type V (wood frame), 67–68
    unclassified, 68–69
United States Fire Administration (USFA)
    cardiac arrest deaths, 1
    potential structural collapse, 81
Unity of Command
    personnel accountability, 21
    purpose of, 17
Unoccupied occupancy, life hazard, 127–128
USFA. *See* United States Fire Administration (USFA)

V
Vacant occupancy, life hazard, 127–128
Vendome Hotel fire, 77
Ventilation
    compartment fire development, 58
    defined, 178
    determining factors for, 178–179
    flashover conditions, 50
    hoselines for safety during, 180–181
    HVAC. *See* Heating, ventilating, and air conditioning (HVAC)
    kerf cuts, 172
    purpose of, 172
    risk mitigation guidelines, 180
    size-up, 180
    tasks, 181
    trenching, 172–173
    vent, enter, and search (VES), 166
    vent for fire, 180
    vent for life, 180
    ventilation-controlled fire, 50–51
Vertical shafts, 61
VES (vent, enter, and search), 166
Victim removal, 167–168
Visual indicators, 132
Void space. *See* Concealed space
Volunteer organizations
    privately owned vehicles, 32
    residential fire deaths, 121
    response time, 32

W
Walls, bearing, 67
Water as structural collapse factor, 80
Water supply
    knowledge of resources, 28, 30
    size-up factors, 129–130
Water tenders, 130
Water-based fire protection system, 129
Weaknesses, personal, 35
Weather
    arrival condition indicators, 131–132
    humidity, 131
    ice, 131
    rain, 131
    response considerations, 29, 103
    snow, 131
    temperature extremes, 131–132
    wind, 132
Wind, impact on operations, 132
Windows as means of egress, 123
Wood as fuel, 56

Y
Your world, 26–29
    components, 26, 27
    life safety, 27
    permanent characteristics, 27–28
    structural information, 27
    water supply, 28
    weather conditions, 29
Yourself, characteristics of, 33–35
    strengths, 33–35
    weaknesses, 35

Index by Nancy Kopper